Seasons with Birds

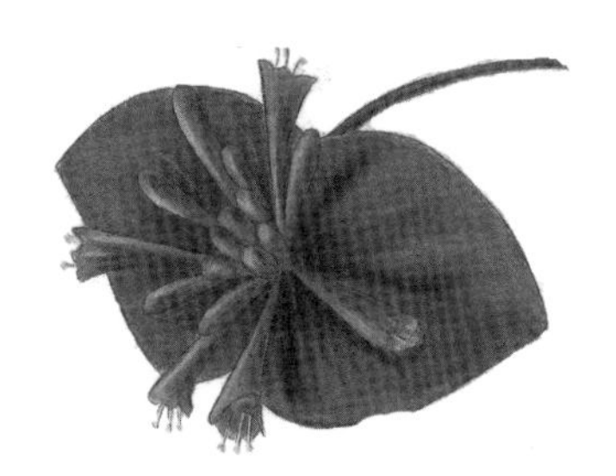

Xantus's Hummingbird

Seasons with Birds

BRUCE WHITTINGTON

Illustrations by LOUCAS RAPTIS

First edition

TouchWood Editions Ltd.
Victoria, BC, Canada
http://www.touchwoodeditions.com

This book is distributed by Heritage House,
#108-17665 66A Avenue, Surrey, BC, Canada, V3S 2A7.

Cover illustration: Loucas Raptis.
Cover and book design: Christine Toller.
This book is set in Adobe Jenson.

TouchWood Editions acknowledges the financial support for its publishing program from The Canada Council for the Arts, the Government of Canada through the Book Publishing Industry Development Program (BPIDP) and the Province of British Columbia through the British Columbia Arts Council.

Printed and bound in Canada by Ray Hignell Services, Inc., Winnipeg, Manitoba.

Library and Archives Canada Cataloguing in Publication
Whittington, Bruce, 1948-
Seasons with birds / Bruce Whittington; illustrations by Loucas Raptis

Includes index.
ISBN 1-894898-21-4
1. Bird watching—British Columbia. 2. Birds—British Columbia.
I. Raptis, Loucas II. Title.

QL685.5.B7W44 2004 598'.072'34711 C2004-905166-0

Contents

Northern Flicker

Introduction

In the beginning, I did not set out to write this book. It was born, I suppose, in May of 1989, when Harold Hosford asked if I would be interested in taking his place as the bird writer at the Victoria *Times Colonist*. Write for a newspaper? Write about birds for a newspaper? For money? Well, I was terrified I was going to wake up.

That was the beginning of "Island Birds," which appeared weekly after that for over 10 years. It was a rewarding time for me, sharing my experiences with readers and hearing their stories about birds too. But one of the most gratifying things I heard from those readers was the frequent question, "When are you going to put out a book?"

With over 500 essays stored on typed pages, yellowing clippings and computer disks of varying vintages, there certainly was a lot of material. But with a growing family and life changes and career changes, there was always too much else on the go, and the book idea stayed on the back burner. Once in a while, though, you are given a window of opportunity, a chance to do something you have wanted to do or might not otherwise be able to do. When I found myself in such a position, I took the opportunity and set about writing this book.

The book is not the collection of columns that first came to mind. Rather, it is a collection of essays about birds and birding. While many of the ideas and events in the book have appeared before, all the essays have been updated, expanded, blended, revised, reworked or rewritten to the extent that I hope past readers will not find themselves thinking they have seen this somewhere before.

Over 300 bird species breed in British Columbia, and the total number of species known to have occurred in the province is approaching 500. Each has its own story and its own preferred habitat, and each habitat has a story too. The essays in this book cannot begin to cover all the birds or all the seasonal avian events we can look for; it is just a sampling, a quick glimpse into the lives of some of these amazing birds. It will take you through a year with birds, a year that starts in September — the month when I have always felt that the year really begins.

The geographic centre of the book is the Georgia Basin, the south coastal region of British Columbia, because that is where I have done most of my birding. However, I have tried to expand the relevance beyond my backyard and my favourite local haunts to other locations in British Columbia and beyond, so I hope readers from elsewhere will find it useful.

I am not an ornithologist or biologist, not a scientist of any sort. I am fortunate to have many friends who are in some way involved in bird research or conservation, and I have learned much from them. I also have a couple of bookcases heavy with volumes about birds, which I have used extensively in the research I have done to write this book.

For many of the essays, I drew extensively on certain published references and they are listed in a selected bibliography at the end of the book. I have done this in part to acknowledge the sources I have used and also to recommend what I have found to be the most useful references.

I don't expect you to know what an "exit pupil" is but you may need to know one day. Since it is difficult to write creatively about exit pupils, I have included a separate appendix for this. Together, the bibliography and the appendix will provide information about two of the most important tools for birders: binoculars (which have exit pupils) and field guides (which do not).

Many of these stories recount experiences I have had on my own, but I have also spent time in the field with many other birders. I have enjoyed their companionship and am grateful to all of them. But in particular, I want to acknowledge Alan

MacLeod, Jan Garnett, Marilyn Lambert and David Stirling, each of whom has played a significant role in shaping this book. Arlene Yaworsky offered many encouraging words and helpful editorial comments. I am grateful also to Loucas Raptis for his remarkable illustrations, and his quiet professionalism and enthusiasm and to Marlyn Horsdal (the Comma Police), whose careful editing made the publishing process easier for a first-time author.

I hope you enjoy these essays. But my greater hope is that they will encourage you to spend more of your own time with the birds, real birds, outside in real trees. Watch them fly, listen to them sing, make your own discoveries about them, bring them into your lives.

Bruce Whittington

Autumn

American Kestrel

September

Bewick's Wren

And Another Year with Birds Begins

For birders (at least for this birder), the best thing about August is that September comes after it. For although the preceding months have been filled with the wonder of another breeding season, after the surge of nesting activity the summer months settle into a routine of predictability.

The songbirds are quiet; their jobs are done, the young fledged and hormones on the wane. The need to sing has gone, along with the urge to court and nest. A few Bewick's Wrens, the Jimmy Cagneys of the bird world, are still demanding attention, but mostly the woods are devoid of song. I can see a few young birds being fed by their parents, and even a few late nesters on the home front; Barn Swallows seem to be more prone to this than most. Immature Great Horned Owls are still screeching at their parents to feed them, but the Rufous Hummingbirds have abandoned my feeders.

September arrives as the busy breeding season winds down and yet, in a very real sense, it is a month of beginnings. With a new generation on the wing, the absolute numbers of birds are as high as they will be all year. For the juveniles, this is the beginning of their lives as birds, lives that will be intense and, for most birds, very short. Some will prepare for a winter in our climate. Their focus will be on survival — finding enough food in a shorter day, to survive a longer and colder night. Others will answer the call of magnetic fields or celestial displays, and navigate to wintering areas in the south. These juveniles, with no prior training, will undertake long migrations to destinations known only to their genes.

This fall migration actually sneaks up on us in July; it's buried under a sultry blanket of summer, but it's there, if we look for it. The shorebirds begin their southward movements, yet it is only short weeks since they laid claim to a patch of Arctic tundra, to lay their eggs and rear their young. After a northward journey of thousands of kilometres, they are on the wing again, back to Central and South America. Some will congregate in the hundreds of thousands to feed en route at Boundary Bay in southwestern British Columbia.

Overhead, perhaps the most visible show of the month will be the annual Turkey Vulture migration over southern Vancouver Island. The hundreds of vultures are often joined by many species of hawks, eagles and falcons. There is usually one movement in late September and another a week or two later.

Along the seashores, the fall's migration marathoners, Common Terns, will move through in a seamless passage of small groups, until there are none to be seen from Victoria's Ogden Point or Whiffin Spit in Sooke. They will complete a figure-of-eight route that takes them back to southern seas, 15,000 kilometres or more from where they were hatched. Undaunted by the weather and sea conditions they face in their travels, they are also harried along the way by other seabirds called jaegers. Pirates of the sea, they outmanoeuvre the terns, which are forced to cough up their hard-won catches of fish and try again.

The terns know that it isn't "fall" everywhere. Here in the Northern Hemisphere, the days are becoming shorter and cooler, and the terns are leaving us to spend the winter on the other side of the equator, where in fact it is not winter at all, but summer. They will return to us when fall descends on the coast of Argentina.

Off the coast of British Columbia a migration of a different sort is taking place in a world of birds often overlooked by backyard naturalists. Here, the ground swell rises and falls to a distant command, and up from the troughs come the shearwaters, black daggers that skim the wavetops. They wheel high on the wind, following every breath of wave-formed updraft and then drop, gliding ever faster through the next troughs.

The Sooty Shearwaters are on the move, drawn to the south not because they have finished breeding, but because they breed in the Southern Hemisphere. They will abandon our fall for an austral spring, to nest on small islands in the south Pacific Ocean. They will stay long enough to see their young fledged and then return to spend most of their year wandering the north Pacific Ocean.

I've come to enjoy birding in the fall more than any other time of year. The seasonal door opens on a hectic level of activity and many challenging birding opportunities. It cannot compete with the feathered brilliance of the early spring, the carolling in our fields and forests, the fecundity of the nesting season. But in the lives of birds, it is a time that is no less critical to their existence. Indeed, it is all the more remarkable, because for many birds it marks their first major passage in life. And so, another year with birds begins.

Putting the World Series in Perspective

Ostriches are big. Hummingbirds are tiny. Great Horned Owls are strong. Turkey Vultures have a great sense of smell. But what is the fastest-flying bird in the world? Here's a hint: It also makes the longest uninterrupted migratory flight on the planet.

The Peregrine Falcon, perhaps? It is probably the fastest bird in a dive, yes, at about 180 kmh, but it falls back in the pack when it comes to level flight. And its migration doesn't fit the pattern. Arctic Tern? It's a good guess, because this species makes what is regarded as the longest round-trip migration annually of any species: as much as 31,000 kilometres. But it stops to rest on land, and for all its elegance in flight, it is not all that fast.

The winner is the Pacific Golden-Plover. Now, before you bring out all your books and magazine clippings, I have to qualify this. In the world of birds, such performances are documented rather less rigorously than home runs in the World Series, and it may not be possible to prove any case conclusively, but consider the facts.

Here is a shorebird built for speed, with long, swept-back wings spanning about 40 centimetres. It is about the shape of an American Kestrel (no mean aviator itself) but it weighs considerably less, about 125 grams. That's about the same as a baseball. It has been clocked in level flight at between 95 and 115 kmh, but experienced shorebird observers have estimated speeds of over 175 kmh. Documented? No, but then, neither is the more celebrated stoop of a Peregrine Falcon. While the falcon relies on its speed for the pursuit of prey, the plover is on the other side of the equation; it uses its speed to escape. Being fast is also a big help with its long-distance migration.

Pacific Golden-Plovers breed in the tundra of Siberia and Alaska. A small number of them migrate along the west coast in the fall, and winter in southern California. The vast majority, though, winter in the islands throughout the Pacific Ocean. Some migrate along the Asian shore to Micronesia, but many make a non-stop journey, from the Pribilof Islands of Alaska to the Hawaiian Islands, a distance of 4,500 kilometres. Cruising at 100 kmh, that's a flight of about two days and two nights. It is an amazing feat of endurance, even from a family known for its strong fliers.

A close cousin of this species is the American Golden-Plover; the two were until recently considered a single species — the Lesser Golden-Plover. Americans, too, are excellent fliers. They migrate from their breeding ground in the western Arctic to the pampas of Argentina but their flight is mostly over land, and they can rest as frequently as necessary.

The Black-bellied Plover is a long-distance migrant as well, wintering along the coast of every continent except Antarctica. It is possible to see any of these three amazing aviators along the coast from British Columbia to California, with most reports of the two golden-plovers coming in the fall. In southern British Columbia, the American seems to be reported more often early in the fall, and the Pacific a little later. My own experience is that the Pacific is the more commonly seen throughout the fall migration on the coast.

Some years ago, I recovered a Pacific Golden-Plover one evening beside a farm reservoir in Central Saanich. The bird was uninjured, but apparently too weak to fly. I took it to a man experienced in these things, and he kept the plover for a few days. Antibiotics cleaned up a minor respiratory infection, and a healthy diet replaced some needed fat reserves on the bird. Then, it was ready to be released.

On a sunny morning at Island View Beach north of Victoria, I carefully pulled the bird from its cage. Its eyes were alert, I could see that, and its body anxious to fly. What I could not see was what was going on in its mind: the urgings of its genes and the navigational calculations whirring along in its tiny brain. I raised my hands and opened them, and those splendid wings lifted the bird aloft and into the next stage of its journey. This plover would probably continue south along the coast, an amazing odyssey by any standard. But I could not help thinking about those other golden-plovers, somewhere now over the Gulf of Alaska, on their way to Hawaii.

That these small birds can store enough energy to power their wings for such a remarkable flight astounds me. I think of a bird the size of a baseball, and I try to imagine the energy and the logistics required to get that baseball from the Pribilofs to Hawaii. But what I find even more compelling is that so many arrive safely in Hawaii. How can they navigate so unerringly, across so vast a sameness of sea, to a few tiny dots of land? How many miss the target? And what happens if they do?

Every year, though, they return — enough of them to produce the next generation. In a world of amazing birds, it is the Pacific Golden-Plover that has taken the laurels in the avian World Series. And when you think about how little they have to work with, and how much they accomplish, it makes baseball seem like a walk in the park.

Listening to the Stories

"I'll go back to Net 13 and open the back three; they'll get the front five on the way to the banding station."

"Okay, I'll start with Net 6 and work your way." One by one, all the nets are unfurled, ties stored safely and the times noted.

There is little more than a rosy promise of sunrise, low on the eastern horizon, but half a dozen volunteers are ready to start the day's work. Half an hour before the sun appears over the Coast Range, the action has begun: A Lincoln's Sparrow has flown into one of the nets, right next to the banding station.

Careful hands quickly remove the bird from the net and place it in a soft cloth bag where it will be safe and quiet. Other birds soon fly into the nets. A Song Sparrow is next, and it already has a metal band on its right leg.

"Looks like we've seen this one before."

"Well, let's get it out of there and on its way."

The mist nets look a little like badminton nets, but they have lengthwise folds in them called "trammels" that the birds softly tumble into after they hit the almost-invisible barrier. The first step in removing a bird from a mist net is to determine which side of the net it went into, and that is not always easy. Then, step by step, the bird is "backed out" by systematically removing the net from the bird. Usually, one leg is freed, and then the other. With the bird cradled in the hand, the volunteer holds the legs between two fingers and a thumb. The other hand is then free to remove the fine netting from one wing, and then the other. Finally, the net is carefully lifted over the head. Then, it's into a bag, and back to the banding station, where it is hung on a hook with the correct net number.

The bander removes each bird in turn from its bag, checking for injuries or stress. The first step is to band the bird. The small, ultra-light aluminum bands are placed on the lower part of one leg, and the number recorded. Then, if the bird escapes before other data are collected, at least it is on record. The bands are supplied in a wide range of sizes for different species, and each carries a unique number. There is also a message to contact the U.S. Fish and Wildlife Service if the band is found (on or off the bird).

Next, the length of the folded wing, or wing chord, is measured and noted. As much as possible, the age and sex of each bird are determined. It is fascinating to learn the differences in, say, the shape of the tail feathers of adult and immature chickadees. Bird banders have developed extremely detailed identification tools that allow them to identify and age birds "in the hand" with a certainty not often possible by birders in the field.

Fat deposits under the skin of the breast are checked to determine if the bird is ready for migration. By gently blowing the feathers apart, the bander can see yellowish fat deposits beneath the parchment skin. Finally, the bird is weighed. It is placed gently, head first, in a sophisticated piece of plumbing pipe with a removable cap on the bottom end. The electronic scale is set to record the weight of the bird, minus the weight of the container. The data are all entered in a log book, along with other significant information (unusual plumage or congenital defects). Some birds are photographed, particularly if they are unusual records for the season or location. Some are photographed because they are exquisitely beautiful.

Then comes the fun part. The cap is removed from the end of the bit of pipe, and the bird is free to fly away. Most fly a short distance, into nearby cover. It takes about 30 to 60 seconds to band and release each bird.

The nets that are used are extremely fine and lightweight, and injuries to the birds are very rare. The birds are removed from the nets quickly and painlessly. Many, like Golden-crowned Sparrows, are quite calm and come out of the net very easily (in 10 to 15 seconds). Others, like chickadees, hang on to anything within reach, and they are best handled by skilled banders, who can remove them most efficiently. The "extractors," volunteers who remove the birds from the nets, check the net lanes every 20 minutes or less, to reduce the time the birds are in the nets, and to minimize the chance of predators finding the birds. Weather conditions are monitored, and the nets are closed if it begins to rain steadily.

This bird banding takes place at the Rocky Point Bird Observatory on Department of National Defence property at Rocky Point, an area that is restricted to the public. Rocky Point is a broad promontory on the extreme southern tip of Vancouver Island, and is a major migratory stop for songbirds. It is one of several migration-monitoring projects being coordinated by the Canadian Wildlife Service, in this case with the cooperation of the Department of National Defence. Each of the projects is supervised by a trained and licensed bander-in-charge.

Over a thousand birds were banded in 1994, the first year of the project. In 2003, in the 10th year of operations at Rocky Point, over 3,700 birds were banded. In the intervening years, very interesting data have been collected. Lincoln's Sparrows, for example, have turned out to be one of the most common migrants, and Dusky Flycatchers, not previously recorded on Vancouver Island, are banded almost annually. Ruby-crowned Kinglets — tiny birds that weigh no more than a quarter — are abundant migrants, crossing the Strait of Juan de Fuca twice a year for as long as they live. Rare captures included two Northern Waterthrushes and Vancouver Island's first Blue-Gray Gnatcatcher. Since 2002, over 500 Northern Saw-whet Owls have been banded in a new nocturnal banding project.

More information will come in as birds banded at Rocky Point are found at other locations. Typically, only 2 or 3 percent will be recovered from elsewhere, but some of these birds may be recaptured if they pass through Rocky Point again.

It is all important information to help us gather a myriad of little details about the life histories of our migratory songbirds. Bird banders have learned, for example, that the males of some species winter in different habitats than the females. That information helps us to determine which habitats must be protected, to ensure the survival of breeding adults. Research has shown that banded birds do not suffer unduly, and their migration and breeding success are not impaired. We routinely saw banded birds feeding within a few feet of our banding table.

The project is continuing, and now the operations are organized by the Rocky Point Bird Observatory Society, a registered charity. There is a paid, licensed master bander in charge at the site, but it is still volunteers who come out to clear the nets, carry out surveys, record data, recharge batteries for lighting on those early mornings, and more.

They know that every bird has a story to tell, a story we need to know if we are to share our home places. Each tiny aluminum band is attached to such a story, and with a little luck, will come back to Rocky Point with a page or two for us to read. It's all in the name of science but, in the end I don't think there's any doubt that these volunteers are really doing it, not for the science, but for the birds.

Civilized Birding

This is civilized birding. You don't have to get up at the crack of dawn. There is no need to slog through mudflats, or peer across scummy sewage lagoons. It isn't necessary to walk for kilometres in order to see more birds. It doesn't matter much if there are 50 other birders, and you can talk above a whisper. This is civilized birding. This is hawkwatching.

The name says it all, but getting a good look at a hawk, any hawk, is not something that happens every day. So, how does one find hawks to watch? Well, the hawks make the job easy for us, as they congregate in the fall, for their flight south.

In eastern North America, one of the premier birding attractions is the fall migration of raptors. Huge numbers of hawks funnel down from their breeding areas across much of northern Canada, and move south along the rocky ridges of the eastern

United States. Just as Point Pelee on Lake Erie is a mecca for birders in the spring, so Hawk Mountain in Pennsylvania is a legendary hotspot in the fall. Birders go in droves to watch the spectacle of hawks moving across the sky all day long, always with more hawks coming to replace them.

The art of identifying hawks in flight at some distance has been refined over the years, largely through the efforts of field observers at places like Hawk Mountain. This refinement was born of necessity, for the traditional field marks are all but invisible when a hawk is far overhead against a bright sky. Birders have come to rely on more holistic identification marks, like the overall shape, the way it holds its wings or how tightly it turns.

It is only fairly recently that birders have become aware of a concentrated movement of raptors over southern Vancouver Island in the last half of September. Groups of from a dozen to 200 are now reported regularly each fall as they move off over the Strait of Juan de Fuca from the southern headlands of Metchosin and East Sooke, west of Victoria. Unlike the major migrations of the east, the most numerous species on the west coast is actually not a hawk but the Turkey Vulture.

There is a sun-warmed rocky knob that I call "Hawk Lookout." To the west is a ridge, and above it the sky is dotted with black. Through binoculars, the speckle of black shapes becomes a flock of Turkey Vultures, about 200 of them, circling against a pile of cumulus clouds.

They soar on wings tilted up in a shallow V, taking advantage of every breath of wind or rising air current to carry them aloft, soaring in a climbing column of birds known as a "kettle."

With the vultures are a few Red-tailed Hawks, looking deceptively small and compact in comparison. They are pale underneath, unlike the distinctive two-toned black and silver-grey wings of the vultures. A Northern Harrier must be identified by its long wings and tail, because from down here there is no sign of the helpful white rump patch that usually identifies the species.

A medium-sized bird moves quickly through one kettle, too energetic for this mundane soaring. It's a Peregrine Falcon, its pointed wings and powerful flight quite distinct. Other kettles produce long-winged American Kestrels and a dark coastal Merlin.

After a time, the vultures move off eastward, toward Rocky Point. Time to pour some coffee, and settle back to watch the other hawks go by. The Red-tails have stayed, and are joined by several Sharp-shinned Hawks, the smallest of the accipiters, and a common migrant. Their shorter wings and longer tails are quite different from the Red-tails' buteo shape, even from hundreds of metres below.

A small flock of Vaux's Swifts roars through just above hat level, and sharp eyes have spotted a pair of high-flying Sandhill Cranes, both migratory species that are often seen at this hawkwatching spot.

On some days Hawk Lookout is shared by only a few birders, but on this weekend day, dozens have arrived to take in the spectacle. Several field trips have converged to bring over 80 people onto the rock, and there is of course no sense of privacy. No matter. Everyone is here for the same reason. Experienced birders identify hawks for novices, and young eyes are helping old eyes find black dots in the sky. Jokes are shared; a vulture is in close, and there is a call for a volunteer to lie down, very still.

A Red-tail puts up obligingly in a snag, and spotting scopes are trained on it. Youngsters are hoisted up to look through the scopes; one spreads his small arms as far as they will go, proclaiming, "It looks *this* big!" Several times, the Red-tails head out over the strait; no, all but one has thought better of it and returned, to try again another day.

The Turkey Vultures seem to spend most of the day over Garibaldi Hill on Rocky Point, but there are always a few moving along the ridge nearby. They sometimes fly up the valley beside us, moving past at eye level. A moulting feather dangling from one wing, the fleshy pink of a carrion-eater's face clearly visible, and the spectacle is a thrill to all, novices and jaded bird-listers alike.

The hawks generally disperse about two o'clock, but the prospect of more birds, and the camaraderie, and the warm September sun, keep many people here long into the afternoon. Perhaps the best part of the hawkwatch is the questions. Why are they here? When will they leave? Are these the same birds that we saw last year? When will they touch down on the Olympic Peninsula? How many go across? How many perish in the attempt?

For many of the stronger fliers, the crossing is routine. For the Turkey Vultures, weak fliers despite their great wings, the 19-kilometre crossing, over chilly water that ruthlessly deflates those thermals, becomes a matter of life and death.

By the time they reach the coast of Washington state, they have scattered, every vulture for itself, and it is almost impossible to determine how many cross successfully. Most probably do, but there are reports of some lucky birds being rescued from the cold water by passing boaters.

Many of the details remain unclear, and the mystery of the migration itself continues to captivate me. It all means I'll have to spend more time hawkwatching, to try to come up with some answers. With birding as civilized as this, I don't think there'll be any shortage of company on the hawkwatch.

Pacific Golden-Plover and Peregrine Falcon

Steller's Jay

October

Five-star Feeders

The tourist season is over for the year. No more looking for Delaware licence plates at the Swartz Bay ferry terminal (it's the only North American species that isn't embossed, and it's rare west of the Rockies). The birds have just about finished passing through, too. If we were going to pick up a tanager, or maybe a Black Swift or Sandhill Crane passing over, it would have happened by now. No, we're moving pretty inexorably out of the summer side of fall birding. And I like this time of year.

I especially like it when I have been organized enough to get my bird feeders scrubbed for the fall while there was sufficient sunshine to dry them out. Although it doesn't hurt to continue feeding through the summer, I sometimes stop, to short-circuit the Brown-headed Cowbirds that show up in spring, and to send the bullying House Sparrows elsewhere while the swallows are nesting.

In other years, the usual reaction I have when the birds of winter return is a mild sort of panic. I remember that I was planning to get the feeders all cleaned out and spruced up. There are a couple of new feeders I put aside in the spring, "waiting for the fall to put them up and see how they work." Uh-oh; fall is here, and the feeders are collecting dust in the — where did I put them so I could find them before the juncos returned? Well, I don't need to panic just yet, but the feeders haven't been cleaned for a while, and now is the time to do it.

Why worry about cleaning feeders, when birds seem to eat whatever they find in the big dirty world out there? When we invite numbers of birds to return to the same little spot, day in and day out, we're giving avian diseases an excellent chance to take hold, and we have to minimize that chance.

Regardless of what they're made of, I clean my feeders with warm soapy water. I dismantle any that I can and do them in the sink, to get all the grunge out of the nooks and crannies.

Wooden feeders are scrubbed to get the dirt out of the grain — I use my toothbrush, and that gives me an excuse to treat myself to a new one. Then I rinse all the feeders and, to minimize the spread of diseases, rinse again with a solution of one part bleach in eight parts water. One more thorough rinse with clean water, let the feeders dry and I'm done.

Now, it's time to load up the feeders again for another winter of window-birding. Shall I get striped sunflower seed or black oil sunflower? Shelled or whole? What about millet — red, white, yellow, or German? Cracked or whole corn? Wheat? Milo? Flax? Thistle? (Somebody actually *picks* thistle seeds?) And what about these blends? Bird Banquet? Sparrow Smorgasbord? For some reason, people think that they have to offer an ever more exotic menu to their avian clientele, as though they were constantly in dread that the feeder critic from the *Junco Times* is just around the corner. Relax. Birds can't read.

There are a couple of simple truths. Many studies have been done to determine what birds like, some by research scientists and some by backyard scientists. The most universally popular seed for birds is sunflower, and the best type of sunflower for wild birds is the small black "oil" sunflower seed. It is a seed that was first used for wild birds because of a failure in the striped sunflower crop. It is eminently suitable, though, because of its high oil content, to fuel the metabolism of a small bird. In addition, the seed is smaller, so it's eaten by many more species, and the kernel inside is proportionally larger. More kernel per seed, more seeds per bag, and it usually costs less than "stripes" as well.

Another staple is millet. Red, white or yellow, it is taken readily by many species, usually more often when sunflower is not available. With these two, you really need nothing else. An ideal mix for a single feeder is a blend of black oil and millet seeds. But it has been my experience that some birds (finches being the worst offenders) will throw out the millet, looking for the sunflower they know is in there. So I offer the two seeds in separate feeders.

What else is good for birds? According to Nanaimo author Bill Merilees, peanut hearts or chopped peanuts are good, but they are a bit expensive. Once the birds get used to these, you may have trouble keeping them interested in the more plebeian fare. Canary seed (which comes from canary grass) is accepted by some species. It has a slippery husk, like flax, and for that reason both of these seeds are suitable for use in gravity-fed tube or hopper-type feeders.

There is a popular specialty seed called "niger," sometimes sold as "thistle." It is in fact an African member of the sunflower family, and somehow the finches have learned to love it. It is expensive (like giving steak to your dog) and the finches are perfectly happy with black oil sunflower, but it's fun for a special treat. You can minimize costly waste by providing it in a small nylon sack manufactured for the purpose.

I have had reasonable success with cracked corn, but it sometimes attracts starlings and blackbirds (which can overwhelm feeders). The same is true of crushed oats and wheat. Milo (sorghum), which is the round seed about the size of a BB pellet found in many mixes, is not popular and neither is canola.

There is a second group of birds that can be attracted, but to a different sort of feed. These are the birds whose diet normally consists of insects and their eggs, and in some cases a variety of other foods. The most common will be the chickadees and nuthatches, with perhaps a wren or a woodpecker for interest. Omnivores like jays, starlings and crows also fit into this category and will be attracted as well.

The main food that is offered to these birds is beef fat, or a mixture of fat and other ingredients. It's usually called suet, but true suet is the hard layer of fat found around the vital organs of cattle. Other types of fat work well too, but the harder they are the better. The commercially prepared cakes of suet are excellent, keeping well even in mild coastal winters.

I've hung chunks of raw suet in plastic-mesh onion bags, but if it is close to a branch, the crows often destroy the bag and haul off the suet in one fell swoop. It's a little better if the bag hangs well below the branch. Some people use rat traps to clamp suet firmly to branches or feeders, and there is a variety of commercial suet cages available.

You can go one step further and prepare a suet-based mix that is so popular that you will spend more time in the kitchen making it than watching the birds. The suet is rendered, or melted, and mixed with other ingredients. I have used a mixture of two-thirds suet and one-third peanut butter with good success. Peanut butter is much better in a suet mix than alone, because otherwise it can stick in the throats of birds and cause suffocation. Bird-food chefs sometimes add bird seed or currants to the mixture as it cools. These mixes can be put into a variety of containers that can be placed outside for the birds. Some people use them in "suet logs," those logs with holes bored in them. Some simply spread the mix on tree bark.

Although many books condone it, I think it's unwise to give birds pan drippings and especially bacon fat. Other ingredients in the fat, especially nitrosamines, might be harmful to birds. These fats also tend to be softer, and if a bird brushes against them, they can foul the bird's plumage and thus ruin its insulating powers. They can also cause infections in feather follicles.

The biggest problem you will run into when offering suet to birds is the inevitable visitation by starlings, which are very fond of suet. One deterrent to their gluttony is a sort of wire cage around the feeder which allows only smaller birds through. This also discourages the starling-sized woodpeckers, of course. To remedy this, try a suet log which is long enough to hang below the wire cage; the woodpeckers can land on it and work their way up to the suet, but the starlings are stymied (he said confidently).

The other solution I have found is to hang the suet so that it is accessible only from underneath. You can do this with a rather crude affair consisting of a square of plywood about 25 to 50 centimetres on each side, hung so that it lies flat, with a wire suet cage fastened in the centre of the underside. Birder and tinkerer Bob Chappell mounts his plastic tub of suet mix upside down on an old cake rack, or refrigerator shelf, so that the only access to the food is from beneath.

Peter Atkin of Victoria designed a feeder that is basically a refinement of the access-from-below theme. Imagine a small shallow tray with a wire-mesh bottom. Then imagine a gable-roof top that fits over this tray. Make it out of western redcedar, and hold it together with a nice piece of sash cord, which also acts as a hanger, and there you have it. It looks great, but does it work? Peter Atkin answers, "Yes," with a twinkle in his eye.

Starlings do a lot of things well, but hanging upside down is not one of them. Neither are they very good at manoeuvring in flight to get themselves upside down so they are pretty much defeated by these devices. One Steller's Jay at my feeder has figured out how to get himself upside down, but the starlings haven't got the hang of it, no matter how much they watch the other birds doing it. They try all sorts of tricks, but to no avail. They may have to resort to evolution in order to take advantage of these suet feeders. More power to them; I'm sure I won't be around to have to face them.

That's your October reminder to get your yard ready for the fall. Because while you may still be denying that the summer is over, my juncos have told me otherwise. So, if you'll excuse me, I'm just going to see what's happening at the feeders.

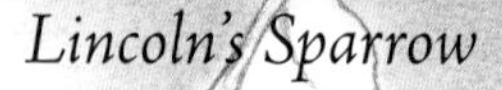

Lincoln's Sparrow

No Problem!

I used to have a standing joke with a friend that went something like this: "What would you like to see today?" he would say as we headed off for a day of birding.

And I would reply, "I'd like to see a Lincoln's Sparrow; I haven't seen one in far too long."

And my friend would snort, "No *problem*! The trees are dripping with them. We're going to get a Lincoln's Sparrow today for sure!"

We didn't find Lincoln's Sparrows on all those outings, of course, and I still count it as a good day when I do find one. What's the big attraction in these little brown sparrows? Well, there are several factors at work.

First of all, to my eye at least, they are uncommonly pretty little birds. They are not showy, parade-through-the-treetops pretty. Rather, they are delicately and subtly marked, such that with every fleeting glimpse, the viewer discovers a little more of the beauty that lies skulking in the undergrowth.

They are always described as being something like Song Sparrows, which is a bit misleading, particularly on the west coast, where our Song Sparrows are quite darkly and heavily streaked. They look pretty much sparrow-like on their upperparts, but have a rusty crown bisected by a narrow grey median stripe. This same grey appears on the sides of the face, and the rust is repeated in a line running back from the eye, down the cheek and back to the base of the bill. Below this, save for a clean white throat, is a lovely, buffy wash across the upper breast. This wash is beautifully streaked with fine lines of rusty brown. In a warm autumn sun, a Lincoln's Sparrow is a delight to the eye.

Another factor in my love affair with Lincoln's Sparrows is that I have a very vivid memory of my first encounter with the species. I was standing at my kitchen window one morning in a pre-caffeine torpor when a small bird appeared in a Garry oak outside. We were separated by no more than a couple of metres and two panes of glass and, even with no field guide I recognized, somewhat incredulously at the time, that I was seeing my first Lincoln's Sparrow. I'm reminded of that meeting almost every time I see one.

But by far the majority of meetings between Lincs and birders are brief and so fraught with foliage as to be almost clandestine. Their close cousins, the Song Sparrows, will jump politely to a visible perch at the slightest squeak or chip or pishing sound, but these guys are very fond of their cover. I am grateful if I get more than a fleeting glimpse.

It's the secretive nature of the bird that gives it some of its appeal; when my patience is rewarded with a better look, the satisfaction is that much greater. And a bird which lives its life so inconspicuously makes me wonder about what it's doing when it's not in Victoria.

Lincoln's Sparrow first became known to science in 1833, when Thomas Lincoln, a 21-year-old assistant to John James Audubon, found one while the two were on a field trip in Labrador. Audubon, after describing the bird as a new species, named it in Lincoln's honour. It is one of three Melospiza sparrows, along with the Song Sparrow and Swamp Sparrow.

Melospiza comes from two Greek words meaning "song" and "finch." While the Swamp Sparrow manages only an unmusical trill, both Song and Lincoln's

Sparrows have quite melodious, bubbling songs. I first heard a Lincoln's sing on a June day in Manning Park, a song vaguely like that of a Song Sparrow, but with some of the rolling qualities of a Purple Finch.

By the end of October, all I'll get from a Lincoln's Sparrow is a variety of *chips* but the bright side of this is that they are here at all. For, while they don't breed in our area, they do pass through in migration, and a few stay to spend the winter. Volunteers at the Rocky Point Bird Observatory in Metchosin have discovered that, despite its reputation as a skulker, the Lincoln's Sparrow is one of the most common migrants on southern Vancouver Island. So I'll be watching for them. I look for them in grassy fields, especially along the edges where there are hedges and brambles and other cover. And I remind myself that they are secretive, and that I'll do well to be patient and wait for the bird to pop up.

And perhaps I'll take a friend along. And I'll tell my friend, "No *problem*! We'll get an American Goldfinch for sure; the trees are dripping with them!" We may not get that goldfinch; but we might just turn up a Lincoln's Sparrow.

The Nicola Valley, October

There are places I am happy to revisit at any time, but the fact is my return trips are usually planned around some highlight of the birding season. There are many other reasons for travel, however, and a fall family visit found me in one of my favourite spots in British Columbia: the Nicola Valley.

It began with the usual rush to catch a ferry, which is enough to temper any enthusiasm for a weekend respite. The prospects of an idyllic getaway faded further with each shudder of the pounding hull, and I watched as the wind tore sheets of spray off the whitecaps. I negotiated the Fraser Valley in teeming rain and, for variety, moved into a blizzard on the Coquihalla Highway. Then, as I descended into Merritt, it looked as though I might have seen the worst of it.

In the morning I awoke on a ranch nestled in the valley of the Nicola River. The sky was brilliant blue, and the air, though cold, wrapped around me, clean and still. I have always loved the ponderosa pines and the openness of their understorey. The deciduous trees had already passed their peak display of fall colour, but the valley was still broadly dappled with yellow aspens and gold cottonwoods. Away from the river, I was caught by the silent spell of the open forestland. No machines combusting internally, spewing pollutants to assault health or hearing. Even the wind was still.

At this time of the year, with the summer visitors gone, I did not expect to encounter much wildlife, so I was glad to be greeted by some of the valley's year-round residents. Clark's Nutcrackers sailed from tree to tree, calling rather softly to each other. To a visitor from the coast, they seemed much less obnoxious than their crow cousins.

A flock of chickadees discovered me as I walked the roadbed of the old Kettle Valley Railway. Nothing new, perhaps, except that these were Mountain Chickadees, with slim white eyebrows highlighting their dark caps.

I was pleasantly surprised by small groups of Townsend's Solitaires. On the coast, I'm accustomed to seeing these thrush relatives occasionally in spring migration, where they are indeed solitary. But I found as many as four together in the Nicola Valley.

According to *The Birds of the Okanagan Valley*, by the three Cannings boys, these solitaires are migratory to some extent, but some are around all year, moving down from higher elevations about the end of October. They eat insects when they are available, and are fond of bees. But when insect prey is reduced by cold weather, they feed on fruits and berries. In this respect, they are like other thrushes.

They are often quiet, but I heard some giving their distinctive call, a whistled *heep* that sounds like a squeaky hinge opened quickly. As the October sun got the better of last night's chill, one of the solitaires began to sing a little. Even in its abbreviated form, the song was ethereal as it carried through an intoxicating stillness.

I was shaken from my reverie by the harsh calls of a family of Black-billed Magpies, the Chilcotin Constabulary; they were escorting a ne'er-do-well Common Raven on its way out of their territory. That done, and still in fine form, they turned their attentions to me, telling me to move along, too.

The weekend, like most, slipped away quickly. Too soon, I left the Nicola, parallelling the lovely river along Highway 8 to Spence's Bridge; I always remember my delight the first time I "discovered" the Nicola one year by taking this road as an alternative route to Kamloops. On this trip, I was surprised by a group of Chukar at the roadside, birds that until then had somehow eluded me on many trips to the B.C. interior. From Spence's Bridge, I was on more familiar ground, and completed a loop by taking the Fraser Canyon to the coast.

These little getaways do much to relax a person and to recharge the batteries, but I am also struck by the enhanced satisfaction that comes from knowing an area in more than one season. Even as short a visit as this left me better acquainted with the Nicola.

So I've learned not to restrict my birding to the "hot" times. A March wedding in Prince George? A conference in Spokane in November? Whatever takes me away from home, I'll have my binocular around my neck, ready to take advantage of the opportunity to do some birding. Even the off-season has its rewards.

What Happened to the Field Marks?

At this time of year, we birders find that our hard-won ability to identify the birds we are seeing around us is taking some hard knocks. What happened to the brilliantly coloured Yellow-rumped Warblers and the neat white wing patches of the Pigeon Guillemots? What happened to the *field marks?*

The answer is at once very simple and amazingly complex. The feathers which the birds were wearing when we figured out how to identify them are gone, replaced by new feathers which in many cases are very different from the old ones. That's the simple part of the answer.

A feather is a remarkable piece of equipment. It is strong, light and resilient, and it sheds water. However, it does not last forever; it needs to be replaced after some months of daily use. Birds replace all their feathers on a regular basis in an orderly process called "moulting."

The process is essentially the same in all birds: each old feather is pushed out of its follicle by its replacement as it grows in. But here we begin to discover the complexities of moulting. Almost all birds replace their feathers completely at least once a year, but many have two complete moults, and most young birds have several plumage changes in their first year.

The down which kept them warm as nestlings is replaced by a juvenal plumage, and this is replaced in the fall by another change into a plumage which will be kept through the winter. The old name for this is "winter plumage" but current use favours "basic plumage."

To complicate things further, some birds shed all their feathers in certain moults, while others retain the old flight feathers. Timing of moults also varies from one species to another.

One of the things that is constant among birds, however, is that the moult is a protracted affair. If it weren't, we'd have backyards full of birds looking like supermarket chickens. Normally, the remiges, or flight feathers, are shed one at a time, in a certain order. In passerines, usually the sequence is as follows: First come the primaries, from innermost to outermost. Next come the tertials, beginning with the outermost. Last to go are the secondaries, in the middle of the wing, with the innermost falling out last. It's quite common to see crows and ravens, for example, flying overhead with symmetrical gaps in both wings; in those gaps are partially grown flight feathers.

The body feathers are shed sequentially as well, but much more quickly. Often you can see individual new feathers that don't fit the pattern of the old; this is particularly visible in shorebirds and gulls.

But even this is too simple. In male ducks, the moult into winter or basic plumage occurs during the nesting season. It is called an "eclipse plumage," and leaves the drakes flightless for about two months. At that time, they undergo what is essentially a very early moult into next year's breeding (or "alternate") plumage.

The most noticeable thing about moulting, however, is that the new plumage can be quite unlike the old. Most birders are aware that young birds don't look the same as their parents; many's the birder who has been fooled by a brown and dowdy Spotted Towhee in juvenal plumage.

The young of the year are around in large numbers in the fall, and trying to assign them to species can really tax your identification skills. Young Cedar Waxwings sport most un-waxwing-like stripes and young Red-tailed Hawks are not red-tailed. Baby Killdeer have one breast stripe instead of two.

But it isn't just the young birds that are confusing. With the need for conspicuous plumage gone as the breeding season ends, some passerines turn into very ordinary little birds. Our warblers, in particular, go from brightly patterned jewels to greenish-yellow birder-bafflers in a very short time. Such is the level of confusion caused by a tree-full of these little menaces that one field guide has given them a whole page of their own: Confusing Fall Warblers.

It's easy to see why birding can be frustrating in the fall. We have to deal with adult birds losing the field marks of spring and adopting not only new, subdued plumages but also much more secretive ways. And we have many young birds (perhaps two or three on average for every pair of adult birds) which are obviously not alternate-plumaged adults, but don't look exactly like adults in their basic plumages either. Add to this the possibility of encountering one of those rare fall vagrants that show up unannounced and you have enough reasons to keep the faint of heart home in bed. But don't give up; if you don't get them this fall, when they return next spring, they'll have field marks again, and they won't stand a chance.

Northern Flicker

November

When is a Woodpecker not a Woodpecker?

Often in November, bird lovers are curious about large brownish birds that have arrived in their backyards. Their descriptions make the identification easy: "Spots on the breast and a large black crescent," "a large white spot on the back or tail," "lots of red on the wings." The birds are Northern Flickers, probably the most common woodpeckers in British Columbia.

In more specific terms, flickers are plastered with field marks. The base colour of the heavily spotted breast is pale buff, and the black crescent is high on the breast, just below a grey face and throat. The crown and nape are a soft brown, which continues down the back, where it is heavily overlaid with dark barring. The male sports a very striking red moustache.

When the flicker takes flight, the effect is dramatic. With its wings spread, it shows a sparkling white patch on the rump. The typical woodpecker wingbeats, several rapid flaps followed by a glide, generate a stroboscopic display of bright salmon red on the underside of wings and tail.

Our flickers are still known to many birders by their former name, Red-shafted Flicker. The name refers to the red feather shafts of the wings and tail. In eastern and northern North America, the Yellow-shafted Flicker replaces it. As its name suggests, it has bright yellow wing and tail linings.

Because the two interbreed freely where their ranges overlap, they were stripped of their status as full species and became races or subspecies of a species now known as the Northern Flicker. A third group, the Gilded Flicker, found in the dry southwestern deserts of Arizona and Baja California, was also included as a race of the Northern Flicker but it has since been elevated to full species status again. Here in southern British Columbia, although they're all called Northern Flickers now, it is an interesting challenge to separate our flickers into subspecies.

While we see primarily Red-shafted birds, hybrids are frequently observed, particularly in winter. These will show a mixture of field marks from the Red-shafted and Yellow-shafted races. Occasionally, a pure Yellow-shafted Flicker will be found.

The differences in the wing and tail are obvious in Yellow-shafted birds, but they also have different markings on their heads. The pattern of brown and grey on the face and crown is reversed, and the male's moustache is black instead of red. There is a bit of red, though, in a bright patch on the nape, known as a "nuchal chevron."

Hybrids can show a wide variety of marks. The flight feathers can range from red-orange to orange-yellow. Some Red-shafts will show red nuchal chevrons, and the occasional bird shows a red moustache on one cheek, and a black one on the other!

As interesting as their plumages are, flickers also provide lots of entertainment in our yards. The common name is thought to have come from its habit of flicking its wings. Some relate it to the birds' nervous bobbing. A theory I like is that the name mimics one of the flicker's calls, which actually sounds a little more like *wicker-wicker-wicker*.

The genus name for the species is Colaptes, from the Greek *kolapto*, "to chisel," and these guys can chisel with the best of them. Often a bird lover finds himself at odds with a flicker that is practising on his house.

However, while the flicker is unmistakably a woodpecker, in some ways it is evolving away from woodpeckerdom. Watch sometime how a flicker sits on a branch. He has two toes forward and two back, just like a Downy Woodpecker, but he prefers to sit across the branch like a perching bird, and not lengthwise like other woodpeckers.

Flickers also spend a lot of time on the ground, which is simply not done if you are a woodpecker. I don't know which came first, but their inclination to forage on the ground fits nicely with their fondness for ants. Flickers are more omnivorous than other woodpeckers. They enjoy typical woodpecker fare like insects and suet, but also eat various seeds. In winter, they often descend on urban backyards to feed on windfall apples.

Northern Flickers are common winter visitors to southern Vancouver Island. Loose flocks of 10 to 30 birds appear in open areas such as Island View Regional Park, near Victoria. In spring, many will leave for breeding areas farther north, but some will stay behind to nest in our area.

You may be able to coax a pair to move into a large nestbox, if you can manage to discourage the starlings. But even if you don't succeed, they'll be back in the winter to brighten a November day.

Infections and Infestations

There is great joy in watching the birds at your feeders, perhaps only a metre or two beyond the glass. It is a real treat to see them close up, to admire details of their plumage and to learn something about their behaviour. Then, sometimes, we see another side of birds' lives.

The first clue is often a lone bird that does not leave the feeder. It rests, hunkered down on its feet, its plumage fluffed against the cold. It looks as despondent and lethargic as you do when you are on the couch with the flu. And there is a good chance it has the same problem. Just like humans, birds get sick. Often they get better; sometimes they don't. It's difficult to know exactly what ails them, but there are clues to some of the diseases seen in birds at feeders.

One of the more common ailments is a viral disease called "avian pox." It appears commonly among House Finches, which will exhibit wartlike growths on their "bare parts": the feet and legs, and the skin around the eyes and base of the bill. Sometimes the symptom is a deformed foot, and for this reason the disease is sometimes called "bumblefoot."

The most advanced cases I have seen have been in Spotted Towhees, although they seem to be afflicted less often overall. It may be that towhees are better able to endure the disease as it advances, while the finches succumb before it is as far along.

The disease is not fatal in itself, but it can interfere with a bird's ability to feed, and impair its vision, making it more vulnerable to attack. Secondary infections can also be a problem when the protuberances are scraped or cut. A second form of the disease affects the bird's mouth and throat, and may restrict swallowing and breathing. Neither form of the disease affects humans.

The virus is sometimes spread by insects such as mosquitoes, but generally it is picked up through patches of abraded skin. Transmission is assisted when birds are in close contact at feeders, and it is important to keep feeders clean, to minimize the risk.

Dirty feeders can also contribute to the spread of other avian diseases. If seed is allowed to become damp and contaminated with fungus or mould, it can cause aspergillosis in birds. Aspergillosis is a fungal disease that attacks the lungs and air sacs, and is ultimately fatal. As the birds feed, they inhale the airborne spores, which lodge in their respiratory systems.

In some years, many birds are found dead near feeders, and it is usually an indication of an outbreak of disease that goes beyond "background levels." For example, the bacterium *Salmonella typhimurium* occurs naturally in passerine birds; most birds survive the illness, as most of us survive the flu. In some years, however, salmonella will kill many more birds, particularly young birds that have not yet developed immunity to the disease.

Salmonella is spread through the feces of afflicted birds, and it is easy to see how birds congregating around feeders can become infected, as they consume seed tainted with feces. Maintaining clean feeders is good practice anyway, but when birds are sick or dying, feeders should be thoroughly cleaned and refilled with clean fresh seed.

The West Nile virus has spread more rapidly than anyone expected since it arrived in North America. It has killed thousands of birds and may ultimately cause serious damage to bird populations. The virus is transmitted by mosquitoes, and for this reason is less of a concern at feeders, except that birds congregate there in numbers. The species of mosquito that are known vectors of the disease breed in small bodies of still water, so keeping bird baths clean will reduce the risk. While the mosquitoes can transmit the disease in turn to humans, it is serious only in people whose health is already compromised. Most people who contract the disease recover with only flu-like symptoms.

When disease rears its ugly head, it usually is not possible to treat the affected birds directly, but there are steps that will help to minimize the spread. Rake away the debris from below your feeders. Some people pour boiling water on the ground to kill some of the pathogens. Paving stones or gravel underneath the feeders makes it easier to keep the area clean. It also helps to move the feeders to new locations from time to time.

It's important not to overlook the bird seed itself. When you buy it, look at it carefully. It should be clean and almost shiny. If it appears dusty, it may have been exposed to damp and mould or mildew.

If you can see stringy stuff inside the bag that looks a bit like cobwebs, avoid it. It is contaminated with insects, and this can be a real problem at home. The webbing is probably the work of a small pest known as the Indian-meal Moth, whose scientific name is larger by far than the creature itself: *Plodia interpunctella*. The moth is small, with a wingspan of about 20 mm. At rest, the forewings fold over the body, and the moth looks grey at the front and coppery or rusty at the back. They seem to sit with the front of their bodies raised. The larvae (little caterpillars) are whitish, and about 20 mm long when fully grown.

Adult moths lay their eggs singly or in small masses in grains, dried fruits and nuts. The larva grows, and then pupates in a silken cocoon, emerging as an adult. The life cycle takes about six weeks, and there may be as many as seven broods in a year.

The presence of these pests in your bird seed is not in itself a big problem. They are not harmful to the birds; the larvae are probably eaten by them. The real problem is in introducing them into your kitchen, where they may infest cereals and baking supplies. Once introduced, they are difficult to eliminate. The larvae can be killed by freezing the seed, or by heating it to 50 degrees Celsius.

Store your bird seed outside if possible. Steel garbage cans are excellent for this; they keep rodents at bay as well. If you have to store your seed inside, smaller sealed containers will work well. Take care to keep your kitchen dry goods well sealed — plastic bags are not good enough. Moths can also be controlled with sticky traps, which attract the male moths with a scent that imitates the chemical compound called a "pheromone" exuded by females. The males enter the trap and land on sticky surfaces inside, being taken effectively out of commission. These pheromone traps are available from pest-control companies and in some houseware departments and feed stores.

With a little housekeeping out of the way, you can settle down for a season of feeder-watching. You can enjoy the birds more and worry less about diseases infecting them, or insects taking over your home.

No, We Don't Get Those Here

Diplomacy does not spring immediately to mind as a trait required by birders. Yet, it is essential to the volunteers who look after the Rare Bird Alert answering machines in many birding communities. A lot of calls about bird sightings can sound pretty amazing. There are usually good reasons why people claim to see these birds. Much of the time too, there are good reasons why they are mistaken. And that's where the diplomacy comes in.

House Sparrows (because they are not real members of the sparrow family) are often not on the sparrow page in the field guide, so feeder watchers sometimes report a Harris's Sparrow in their backyard, black throat and all. It's an easy mistake to make, and it's a tricky job to correct observers without making them feel foolish for their misidentification.

Snow Buntings sometimes get reported. At the feeder. On top of the telephone pole. Eating suet. About the size of a chickadee. About the size of a crow. While Snow Buntings are common in winter in some parts of North America, they are pretty scarce around here, and are usually seen only in migration, in open fields, especially along the coast. Often, though, observers see birds of other species that are exhibiting various degrees of albinism. They have white patches, or an overall paler cast, and they may superficially resemble winter Snow Buntings.

It seems, too, that the bird at the feeder can only be the vagrant finch, the non-migratory dove, the almost-extinct warbler. More than one caller has been politely told, "No, we don't get those here."

It was with some unspoken doubt, then, that the Victoria Rare Bird Alert reported a sighting of a species that had never been documented before in Victoria. The RBA noted that the sighting was backed by a good description: Smaller than an American Robin; all grey; darker wings and tail; black cap, and a distinctive rusty patch on the crissum (the underparts at the base of the tail). Hmmm — Gray Catbird — nothing else could come close.

And sure enough, birders tracked it down and confirmed the identification, correctly made by a careful novice. It took a little time, because Gray Catbirds are notorious skulkers, especially in winter. The birders were also taking care not to harass the bird, either by pursuing it in its adopted haunts at Rithet's Bog in Victoria, or by playing taped calls to draw it out. But patience paid off, photographs were taken and the species was added to the Victoria Natural History Society bird checklist.

Catbirds are not all that exotic. They're fairly common in the Okanagan Valley, and are seen regularly in the Fraser Valley. There are several records for Vancouver Island, and a few from nearby islands. That makes sense, because catbirds motor along at a vulnerable 25 kmh or so, and they like to stay low and close to cover. It's possible that all these catbirds island-hopped across farther north, and then made their way south.

This particular bird had a long way to go. Most catbirds winter in a narrow strip along the Gulf of Mexico, some as far south as Panama. Even if this catbird found its way to the coast, the Strait of Juan de Fuca could discourage it from attempting a crossing. That would mean spending the winter here and, El Niño or no, it's going to be longer and colder than it would have been in Galveston.

It always hits home to me that vagrants like this carry no real biological significance; they are aberrations, misfits. But there is an undeniable magic in looking at an individual bird and speculating on its genetic programming, its travels and, even more compelling, its future.

Did it settle in at Rithet's Bog? It certainly was taking advantage of a good crop of hawthorn fruit. Did it squeak through a milder-than-usual coastal winter? If it survived, where did its migration computer take it next spring?

It raises other questions, too. Was this really the first Gray Catbird to appear in Victoria? What are the chances that others have arrived, unseen? I'd say pretty good. And how many people have reported Gray Catbirds, but could get nobody to believe them? "No, we don't get those here; you probably saw a towhee." One of the nice things about a confirmed sighting such as this is that it sets a precedent; from now on, catbird sightings will have to be given more credence. We can be thankful for the skill of a new birder, and her courage in reporting a "first Victoria record." And we can also be thankful for the pluck and stamina of one lonely grey bird, in a hawthorn thicket at Rithet's Bog.

Xantus's Hummingbird

The Enigma of Xantus

If birding does not already have enough to hold your attention, we can add enigma as one of its appeals.

Some years ago, a hummingbird showed up in a backyard in Gibson's, on the Sunshine Coast of British Columbia. That's to be expected, except that it was November, and this bird turned out to be a Xantus's Hummingbird, a species native only to Baja California. Non-migratory, it has been recorded only once north of there, in California. So why Gibson's? The question of the origin of a vagrant like this can generate heated discussions, and this case will keep the discussions warm for some time.

What is the likelihood of a Xantus's Hummingbird showing up in B.C.? Slim to nil. But what about El Niño, and those tropical storms in Baja? Well, maybe. What about the Tropical Kingbirds that appear in some autumns, and that White-winged Dove that came to Tofino? Couldn't they all have been blown north the same way?

It's possible, of course. But there was also a rumour going the rounds that a bird fancier in the Lower Mainland lost some illegally kept birds, and that he is believed to have had some of these "vagrant" species. Will we ever know the truth for sure?

If this Xantus's Hummingbird is enigmatic, it has in some ways come by the trait honestly. The first specimen was reported to science in 1859 by Jonas Xantus, who was under contract to monitor a U.S. Coast Guard tide gauge at Cabo San Lucas, in Baja California. Enigma might have been Xantus's middle name.

Xantus had been educated in his native Hungary, and enlisted in the Hungarian army. He was captured when Hungary fell to Austria in 1849, but escaped and fled to the United States. His early years there are little known, but it seems he wandered in search of employment he felt was suitably prestigious. In desperation, he finally enlisted again, in the U.S. Army.

He befriended the ornithologists William Hammond and Spencer Baird, and through their efforts was posted to Fort Téjon in southern California. Here, in addition to his regular duties, he collected tremendous numbers of bird and animal specimens and sent them to the Smithsonian Institution. When he tired of this posting, he jumped at the opportunity to leave the army and move to Baja California. He continued his collecting at a frenetic pace, cataloguing 130 species of birds, as well as many other terrestrial and marine organisms.

Xantus earned an enviable reputation for the breadth of his collections and the quality of his specimens. He was a vain man, however, and the recognition was never enough. Much of his writing has subsequently been shown to be plagiarism, even down to his use of other artists' illustrations. He adopted the name "de Vesey," which he added to Xantus, because he thought it would command more respect. In the best-known illustration of him, Xantus is portrayed in the uniform of an officer of the U.S. Navy, which he never joined.

He moved from position to position in North America, but ultimately he returned to his homeland. He first sent gifts of specimens to the Hungarian National Museum, and much written material to the press. Welcomed, then, as an eminent naturalist, he still could not resist lecturing about places to which he had never been, and people with whom he had never travelled.

He succeeded finally in obtaining work as a naturalist, and subsequently finished his career in the ethnographical department of the National Museum. A first marriage produced a son in 1874, but ended in divorce. He married again and lived the rest of his life in Hungary. He never recovered fully from a bout of pneumonia in 1894, and died that December at the age of 69.

Although Xantus's indiscretions were pointed out as early as the 1860s by members of the American-Hungarian community, he escaped censure until the 1940s, when clear proof of his questionable ethics was presented. At that same time, the novelist John Steinbeck discovered another of Xantus's legacies; a whole lot of Xantus descendants among the natives in the hills back of Cabo San Lucas.

I don't know which is more enigmatic, the hummingbird in Gibson's which should be in Baja, or the man whose name it bears — liar, plagiarist and accomplished naturalist.

Ahar-a-lik

As an autumn snow blanketed the Saanich Peninsula on southern Vancouver Island, the birding opportunities looked poor. Ten centimetres of the stuff lay everywhere, and gusty winds blew ice pellets into faces foolish enough to venture out.

It seemed the best thing was to do the few errands that could not be avoided, and perhaps sneak in a few birds from the comfort of the car. Birding like this is pretty limited, but there is one place you can always find birds in a snowstorm: salt water. I drove out onto the government wharf at the foot of Beacon Avenue in Sidney, where the water is only two or three metres away (straight down) and the view is unobstructed.

Several other cars were there too, but birders they were not. They were watching a west-coast ritual — sailboats of all sizes, tacking back and forth, missing each other only narrowly, as they manoeuvred to cross the start line exactly as the horn sounded to begin the race. Every open cockpit held three or four sailors, fully exposed to the elements. While most wore bright yellow survival suits, I couldn't help but think that they were either hardier or crazier than birders.

As the last horn sounded, the tangle of fibreglass and Dacron arranged itself into a line of boats, all scudding south toward the first marker. And across the empty water they left behind, two Long-tailed Ducks dodged their way northward. As they entered my field of view, they had me once again; these ducks of our winters always get my heart going a little faster.

Thousands of sea ducks spend the winter in the productive waters of the southern Strait of Georgia, but the Long-tailed Duck somehow stands apart. One of the reasons is its erratic flight. It tilts back and forth, alternately flashing its dark upperparts and white lower parts, looking remarkably like a shorebird as it tips and turns.

The male is unique among the ducks in having three separate and distinct plumages. We most often see the winter or basic plumage. The head is white, with a dark smudge on the cheek. The wings and breast are dark, and the white of the belly is complemented with patches of white on the scapulars — what we think of as shoulders. As spring approaches, he loses these white scapulars and becomes brownish above. His face pattern reverses, the head becoming dark with a white cheek patch. In both seasons, he sports the long, elegant tail plumes that give him his common name. As in most ducks, the males also moult into a female-like eclipse plumage, which they wear after losing their breeding or alternate plumage, and before their new basic plumage comes in. Females are less clearly marked, but the pattern of colour on their heads changes with the seasons as it does in the males. Their tails are elegant, too, if not quite so long as in the males

The Long-tailed Duck in winter is a noisy, social duck. It's usually found in loose flocks of up to 50 birds. Its garrulous nature is thought to be the root of its former common name: Oldsquaw. Other folk names include Old Wife, Old Nanny and Old Molly. The name Long-tailed Duck has been accepted in recent years in North America to match its long-time usage in Britain and Europe.

The calls of the Long-tails carry widely across the water, a gabbling that's difficult to transliterate in English. It's often written as *ow-owdle-ow*. A much better description, I think, is found in the name given to the bird by the Inuit of the Mackenzie delta: *ahar-a-lik*.

In March and April, the Long-tailed Ducks move north to the tundra for breeding. Their breeding range covers much of the Canadian Arctic, including the Arctic islands. There are two breeding records, 60 years apart, from the northwestern corner of British Columbia. Both records are far from the normal range of the Long-tailed Duck, and this raises the question of whether there are other disjunct populations.

They nest on the ground, usually not far from water, and often in close association with Arctic Terns. The terns are aggressive defenders of their own nests, and the Long-tails take advantage of this protection from predators.

The young birds grow very quickly, and several broods will join together to form what is known as a crêche. All the youngsters are guarded, usually by non-breeding adults; some crêches have been recorded as including 135 ducklings! This arrangement for the care of the young has, on the surface at least, elements of altruism, but researchers believe that it is simply a strategy of strength in numbers.

All these details come together to make a Long-tailed Duck what it is. But somehow, the whole is more than the sum of the parts and any day is a good day when Ahar-a-lik zigzags through my field of view.

Long-tailed Duck

A Page from a Birder's Diary

September 9th

I've spent a lot of time on this, reshingling the walls of this old house. It's tedious, but not unpleasant: sort of an absorbing tedium. These gable ends are full of character, but at a price. Everything has to be done from a scaffold now. I've quickly learned that working on a scaffold means many wasted trips up and down, and so everything I need for a few hours' work goes up with me right at the beginning. Tools, nails, shingles, water, binocular.

There is quite a good view up here, above the garden, and a chance perhaps to see a few birds. A little later in the fall, there will be hawk-dots among the clouds to squint at, but for now at least, well, the task at hand is interrupted only occasionally by a marauding band of chickadees.

Today, though, as I look down on the top of an old apple tree, another sound comes to me. It's softer than the chickadees, and almost musical. Bushtits? Haven't seen them for a long time. But sure enough, soon the branches of the tree are alive with the tiny grey bodies.

A flock of Bushtits is an addictive thing. At first, I simply watch them, and then maybe try to count them, but I soon give that up as a hopeless task. Then there comes the luxury of settling on an individual bird, and following it on its way. It will land in the tree and immediately it is feeding. Out to the tips of the branches it goes, investigating all the surfaces of the leaves. If the vantage point is not quite right, it will hang upside down for a better view.

For a bird that is so distinctive by its size, the Bushtit is otherwise very nondescript. The upperparts are variable grey, with a browner crown, and the underparts are paler. The tail is long, the bill is short, and that's it. Nothing fancy about these guys. Males and females are similar, except that males have whiter throats, and adult females have pale instead of dark eyes; there, I see one now.

Once the Bushtits appear in the fall, they seem to come around on a regular basis. They go through the yard, from currant bush to apple tree to ornamental shrub, picking insects from the foliage.

Here on Vancouver Island we haven't always enjoyed their company. The Bushtit is widespread in the southwestern United States, but farther north its range narrows to a strip along the coast. In British Columbia, it was formerly restricted to the Lower Mainland, but since the 1940s it has been found on Vancouver Island, with the first nest at Victoria recorded in 1945.

They have expanded north along the east-coast lowlands, commonly to Courtenay and occasionally to Campbell River, and I've seen them also in the Alberni Valley, evidence of an interesting extension of this biotic zone, west via the low passes. They're in Port Renfrew in small numbers too, possibly working their way west in the narrow strip of suitable habitat that parallels the highway. They are still restricted enough to be a west-coast specialty, though, and visiting birders often have Bushtit on their Canadian "want list."

These busy flocks will stay together throughout the fall and winter, when seeing Bushtits becomes a bit of an all-or-nothing affair. But while Bushtits are gregarious for most of the year, they are not colonial nesters. In February or March, the flocks will unravel into pairs, and they will proceed with courtship and mating.

I wonder where these birds nested. People often find the nests of Bushtits, and it is no wonder. They are frequently situated in open areas, often over a path through the woods and routinely about two to three metres above the ground.

Although the Bushtit is one of the smallest of North American birds, its nest is a marvel of engineering and industry. It looks like nothing so much as a sock — an old grey woollen work sock, suspended from a branch or crotch and about 20 to 30 centimetres in length. It is neatly constructed of mosses, lichens and grasses, bound together with spider silk. There is an opening near the top, but often the true entrance to the nest is concealed, lying slightly below this upper opening.

Five to seven eggs are laid in this nest and are incubated by both sexes. It is interesting that both sexes also roost overnight in the nest during the incubation period. The young hatch in about 12 days. They develop quickly on a diet of insects and spiders, and are ready to leave the nest about two weeks later. Despite the apparent security of the Bushtit's pendulous nest, there are a few recorded instances of parasitism by Brown-headed Cowbirds, which lay their eggs in the nests of other birds. (I have a hard time imagining a female cowbird backing up to the opening of a Bushtit nest, and successfully offloading her cargo.) When the young are out of the nest, the family group forages together and, as other families are fledged, the groups meld to form larger feeding flocks.

Bushtits seem very happy in this garden, stopping by regularly on their neighbourhood rounds. It does my heart good to watch them cleaning up insects and eggs on the fruit trees. They seem especially fond of the seed heads of Ocean Spray, and I've watched them feeding also on the seed heads of *Buddleja*, or Butterfly Bush.

As insectivores, they are easily attracted to the suet I put out. A little ball of suet hanging in a metal cage or nylon-mesh onion bag will one day be festooned with Bushtits. They are quite confiding and can often be called in very close by pishing or squeaking. On a cold winter day, I might find myself brought face to face with a tiny dustball of a bird, which somehow manages to feed its little furnace enough to keep the fire going until tomorrow.

For now, though, the going is a little easier for the Bushtits. There is food aplenty, and they seem to know it. Unless it's my imagination, I think they are just full of the joy of being Bushtits, and that's worth putting down the hammer for a look any old day.

Winter

Belted Kingfisher

Western Tanager

December

A Hundred Million Birds and One Siren

Not long after the shots were heard, two headlights could be seen, coming straight at them. As they stepped aside, the car shot past, spun 180 degrees, and came at them again. It isn't often a bird writer gets to begin like this. But then, this sort of thing doesn't often happen on a Christmas Bird Count.

"They" were the late David Pearce and company, Victoria birders out in the wee hours of December 18th, 1993, in search of owls. The car was a police cruiser, out in the same wee hours in search of the source of gunshots which were heard in the vicinity of Thetis Lake, near Victoria. The police set off in the direction the birders pointed, and the birders continued with their count.

The Christmas Bird Count is a very long-standing tradition in North American birding. The first count was organized in 1900 in Massachusetts, by ornithologist Frank Chapman. He felt that birdwatching had a future as a sport and organized the count to protest the so-called "side hunts." These hunts were held on Christmas Day, and involved two sides competing to see which could shoot the most birds (of any species) in an outing. Times have changed.

Over a hundred years later, Christmas Bird Counts are held in about 1,800 North American locations, involving an estimated 50,000 participants, and tallying well over a hundred million birds — seen, not shot. The value of the data that have been collected is immeasurable.

Counts are conducted in a circle with a diameter of 24 kilometres. The circumference is carefully drafted to include as many habitats as possible, and once determined, the centre does not change. The date of the count must fall between December 14th and January 5th. Most count organizers schedule their counts on the same day each year, such as "the first Saturday in the count period."

A smaller, dedicated (less complimentary adjectives have been used) group of birders is in the field early, shortly after midnight, to locate and tally as many owls as possible in the count circle.

Most participants, though, start when there is enough light to see well by and continue until that light fails. Many groups stop for a bit of socializing and a mid-day recap over lunch; some eat on the run. A few are treated to seasonal libations at certain jealously guarded birding hotspots along their routes.

This is a birding event in which birders of all skill levels can participate. Experienced birders may be given an area of their own to cover. Less experienced birders will go as part of a team; they can help in spotting and recording sightings, and can also learn from those who have been birding longer. Another, less celebrated group of participants, known as feeder-watchers, spends the day at home, to record the birds that turn up there.

The birders come in all shapes and sizes. Ian McTaggart-Cowan, the dean of B.C. ornithology, has participated well into his ninth decade, and many a pink-cheeked bundle has birded the count from the comfort of an infant backpack. Coastal counts generally find up to 200 birders in the field for each circle, regardless of weather. Edmonton holds the North American record, though, for the highest number of participants in a Christmas Bird Count, boosting its total to over 1,200 with a large contingent of warm, cozy feeder-watchers.

Every count is different, of course, and every count has its tales of amazing rarities and birds that got away. Most pigeons, for example, grow up never knowing what it's like to be appreciated by birders, and that's because most birders don't see a lot to appreciate in pigeons. After all, they're common, and not native to North America, and their status as wild birds is always somewhat in doubt. But on one Victoria Christmas

Count, there was a glimmer of hope in pigeondom. For a flock of about 30 birds along the east side of Prospect Lake Road, immortality was but a few flaps away.

On the west side of the road stood Mike McGrenere, birding at the edge of his area of the count circle. This area, known as the Central Highlands, is primarily heavily forested, and has never recorded a Rock Pigeon (as these birds are more correctly known) on the count. Mike was, understandably, eyeing these pigeons with, shall we say, desire. One might even go so far as to call it lust.

And then, the curtain rose on every birder's dream: a Sharp-shinned Hawk blasted out of the trees behind the Rock Pigeons, which all took flight, some of them heading straight for Mike's area. Mike held his breath. And then, the curtain went up on every birder's nightmare: they all, in perfect pigeon protocol, broke and banked hard, and headed back the other way. Not a wingtip entered the Central Highlands.

Nanaimo birders caused a bit of a stir in 2000 when they found a Rose-breasted Grosbeak a long way from home, on their annual count. A Lewis' Woodpecker on the 1994 Duncan count was a bittersweet reminder that the species no longer nests on Vancouver Island. In 2002, kayakers birding Prospect Lake on the Victoria count were stunned to find a Red Phalarope; large numbers of these little shorebirds had abandoned their offshore haunts of winter to delight birders on the Victoria and Sooke counts.

When all the birds are tallied, coastal British Columbia always comes out tops in Canada for the highest number of species on a Christmas Bird Count. Victoria set the Canadian record at 152, but Ladner now shares this honour. Vancouver also does well, and Duncan and Nanaimo are two counts to watch.

The species count is fun for us birders, but one of the most important benefits of the Christmas Bird Counts is the mass of data about bird distribution that is collected. Is 1,098 Trumpeter Swans in Courtenay higher or lower than normal? How long ago was the last big irruption of Red Crossbills? Is 4,500 American Robins over Scafe Hill unusual? We have pretty good answers to these questions and more, and it's all thanks to the contributions of hundreds of people who put a little extra effort into a seasonal birding tradition — with or without sirens.

The Halcyon Days

The planet has turned and tipped until finally it has given us the shortest day of the year. There is hope in the fact that the gloom of winter can only get better, but for now we have to contend with serious restrictions on sunlight. These are the halcyon days. No, not those languid days of summer: these days, right around the winter solstice.

In Greek mythology, Alcyon was the daughter of Aeolus, god of the wind. When her husband, Ceryx, drowned at sea, Alcyon was so distraught that she jumped into the sea herself to join him, and both were transformed by the gods into kingfishers.

Kingfishers were thought to build floating nests at sea, and the gods dictated that the seas should be calm during this time so that Alcyon could nest successfully. The Roman scholar, Pliny, recorded that kingfishers indeed nest in the sea, in the winter, "in the halcyon days, wherein the sea is calm." The legend is honoured in the scientific name of our Belted Kingfisher, *Ceryle alcyon*. Like Ceryx and Alcyon, kingfishers pair for life. However, the similarity between the legend and the kingfisher ends there.

Belted Kingfishers nest not at sea but in burrows that typically are excavated in vertical banks, which may be natural or man-made. The horizontal burrows are usually about a metre deep, but may extend into the bank as much as three metres. Both male and female excavate the burrow, using their feet. The feet are uniquely adapted for this: the front toes are joined along part of their length, which aids in the digging.

The nest, as you might have guessed by now, does not float, and neither is any material used in its construction. Six or seven eggs are laid on the floor of the burrow and are incubated by both sexes. When the young have hatched, they are fed by both parents.

Belted Kingfishers eat small fish almost exclusively. They may locate these from an exposed perch, or in flight. Kingfishers, like hummingbirds, are unusual in their ability to hover, and they hang over their prey until it is time to strike. Then they plummet into the water and snatch the fish in their bills. Their eyes are protected as they hit the water by the nictitating membrane, a sort of eyelid that closes over the eye. They usually prefer fairly shallow water, and rely on their momentum to propel them to their prey; then, they put their wings to work, to lift them free of the water. They fly away to a nearby perch, where they usually beat the fish into submission before swallowing it.

Not all kingfishers fish, though. Some members of the family lead more terrestrial lives, eating insects and small vertebrates. The Laughing Kookaburra of Australia is one example. As a family, kingfishers range from tiny to large, and a book about kingfishers is a riot of vivid colours. The Belted Kingfisher is the only species found in Canada, and is rather subdued compared with some of its cousins in Africa and South America. In an unusual twist among birds, the female of the species has an added splash of colour. Where the male wears a belt of blue across his breast, she has an additional belt of rusty brown. The colours are similar to those used in heritage houses, and I wonder if the idea was inspired by some long-ago kingfisher.

Our kingfishers do not migrate. They live solitary lives during these winter months, but they will rejoin their mates in the spring. They will wait almost until the next solstice, in June, to raise their families. And while those may not be the halcyon days, they will be the *alcyon* days.

Talking Turkey

I was brought up in a white Anglo-Saxon Protestant milieu, so Christmas plays a role in my life. There are a lot of birds that might come to mind when I think of Christmas: French hens (whatever they are) and turtle-doves are pretty obvious, but probably the best known is the Christmas Turkey.

Now, the Christmas Turkey is not a full species; it's merely a subspecies which has a particular destination programmed into it: it's going to end up on a lot of tables on Christmas Day.

There must be a jillion fewer turkeys afoot in the world on Christmas Day than there were a couple of weeks earlier. But where did they come from? We all know that they were raised on farms of some sort, but what was a turkey like before we could buy it wrapped in plastic instead of feathers? Like our domestic chickens, which are all descended from the jungle fowl of the Indian sub-continent, turkey lineage can be traced to wild birds. That ancestor is the Wild Turkey.

Wild Turkeys were historically widespread throughout Mexico, the eastern and southern United States and north to Canada, but their numbers declined as a result of habitat loss. They occur naturally nowhere else in the world. There are only two species in the turkey family, and the second is even more restricted in its range.

The other species is called the Ocellated Turkey of Mexico's Yucatan Peninsula, and it is a more striking bird. The bare skin on its head is an arresting bright blue. There are multi-coloured warty protuberances on the head as well, but no fleshy snood on top of the head, or "beard" around the bill. Perhaps the most striking feature of the Ocellated Turkey is the large eye-spots on the tips of the tail feathers; the Wild Turkey has a more subdued tail band instead.

Turkeys are large, ground-dwelling birds, and they are primarily vegetarian. Add to this the fact that they are a good food source and you can see why people wanted to have a few turkeys in the yard instead of having to chase them around out in the bush. Turkeys (and it seems only Wild Turkeys) were first domesticated by the natives of Mexico, in pre-Columbian times. Along with other early treasures, they were brought to Europe and raised there. At some point, people seem to have forgotten where the birds originated, because they became known as "Turkey hens" and "Turkey cocks." It is assumed that people thought the birds had come from Turkey. The terms for the different sexes were gradually dropped, and the misnomer has been with us since.

Turkeys were later given a free ride back to North America as farm animals arriving with early settlers. People must have been chagrined to find them running wild over much of what is now the eastern United States, after having given room on their tiny ships to their breed stock. The settlers soon discovered, too, that the subspecies they encountered in their new homes was larger than their descendants of the Mexican race, so they set about crossing the larger birds with their own stock in an effort to increase the size of the table birds. That trend has now reversed somewhat, with smaller turkeys more in demand these days.

Canada never had very many Wild Turkeys. Historically there were fair numbers in the woods of southern Ontario, but they fluctuated quite wildly depending on the severity of the winters. The species succumbed to pressures of habitat loss around 1902.

Like many gallinaceous birds, Wild Turkeys have been introduced in many parts of North America, particularly in the west where no birds existed previously. They can be found in pockets in Washington state and in southern B.C.

According to the late "Davey" Davidson of Victoria, Wild Turkeys were introduced to Sidney Island for a private hunt. He reported the introduction in 1966 as having taken place "a few years ago" and indicated that the birds were doing well. Keith Taylor notes the introduction in his *Birder's Guide to Vancouver Island* but points out that the birds were always dependent on feeders for survival.

It's unlikely that we'll ever see a Wild Turkey in the wild on Vancouver Island; the closest we'll get will be that familiar carcass on the platter, a long way from the wild, and an even longer way from pre-Columbian Mexico. But it's appropriate that this tradition has pretty deep roots right here in North America.

And when the dust has settled from the celebrations of the season, it's nice to get out for a bit of fresh air and watch some birds. For me, it's one of the best ways to deal with too much turkey!

Western Tanagers and the Double Mocha Latte

Our vehicle slows as it enters the little village of La Bajada, compelled to do so by the large round cobbles that make up the main street. It is a village typical of the hills of the state of Nayarit, in western Mexico, with chickens at the roadside, and the sound of brooms sweeping away the dust that settled overnight.

It takes only a few bouncy minutes for us to pass through La Bajada and pull over at a wide spot in the roadway. We are now in an open woodland, with an understorey of shrubs and a canopy of tall deciduous trees.

Western Tanager

Our group is out of the vehicle quickly and is soon greeted noisily by Golden-cheeked Woodpeckers and Orange-fronted Parakeets. Arriving just after dawn, we have found the birds already on the move. There is a mixed flock here, typical of subtropical forest habitats. Rufous-backed Robins work the understorey, much like their North American cousins. Overhead, Greyish Saltators have joined a group of Streak-backed Orioles feeding on nectar in a flowering tree.

As the flock moves, we follow it up the rocky trail. Beneath our feet are stones laid hundreds of years ago by the Spaniards, to carry their wagons of plunder down to the old port city of San Blas. The old road, known as the Camino Réale, is bordered by shrubs with glossy green leaves, and berries of green and red. These are coffee plants, scattered through the forest, and they are cared for and harvested just like any other crop.

This is coffee as it has been traditionally grown. Like many plants, these coffee bushes grow best in the shade of taller trees. The berries, each with two developing beans, mature slowly, and are picked, one by one, as they ripen to a bright red.

Overhead, in the shade-producing canopy, there are birds. Mexican Parrotlets, sparrow-sized jewels with voices like peacocks, vanish by the dozens in the foliage. A pair of Gray-collared Becards has joined our mixed flock, and now there is a tanager, yellow with black wings and a red head: It's a Western Tanager, the same bird that moves through backyards in southern British Columbia in the middle of May. There are familiar warblers, too, Orange-crowns and Black-throated Grays, feeding to prepare for their long flights north to British Columbia to breed.

Ahead, a Russet-crowned Motmot flies silently to a shadowy perch, its racket-like tail feathers undulating behind. Rufous-bellied Chachalacas set up a cackling which carries across the open forest. Low down, gleaning insects from among the coffee plants, are more warblers — Wilson's and Nashvilles — all bound for points north in a few months.

As the Camino Réale crests a hill, at about 450 metres, the sun brightens. The canopy disappears, and we are in a banana plantation. Suddenly, the birds are gone. Or, rather, they are now behind us, where the forest lingers at the edge of the grove.

It is a startling change. We North Americans love bananas, but banana plantations make very poor bird habitat. On the other hand, coffee — that warm, dark friend of our mornings — grows in the shade of a forest which is home to hundreds of species of birds, both residents and North American breeders which winter there.

This eminently satisfying situation is rapidly changing. In many coffee-growing areas, new strains that are sun-tolerant are being planted. With the increased sunlight, they produce more rapidly and more abundantly, and are more profitable. Sun-grown coffee is a technological marvel. Or is it?

Sun-grown coffee plantations fall into the category of ecological monocultures. These habitats, which consist of one species only, or a few at best, support very low wildlife diversity. Research in Mexico and Colombia has found only 10 percent of the bird species in sun-grown coffee plantations, compared to plantations that produce coffee in the shade. The tremendous diversity of bird species in shade-grown coffee plantations represents both resident populations and those species known as the neotropical migrants, birds that North Americans like to think of as "ours." It is another lesson in the sterility of monoculture habitats, rivalled in North America by tree plantations, industrial agriculture and The Lawn.

Sun-grown coffee plants produce as though they are on steroids. They begin to bear fruit at a younger age, and have a shorter productive life, than shade coffee plants. The increased productivity puts a greater demand on the soil, soil that is not restored by the natural humus of a forest ecosystem. As a result, it requires fertilization, with all its consequent problems of runoff and leaching. Sun-grown coffee plantations have no canopy of trees to support a population of insect-eating birds, so insect pests flourish, and the crops must be intensively managed with pesticides. Some research has shown that these sun-grown coffees are tainted with levels of chemical toxins many times higher than permitted levels.

In northern Central America, coffee plantations make up about 44 percent of the cropland currently in production. If that land continues to be converted to sun-grown coffees, it will mean a devastating loss of habitat for the resident and neotropical migrant songbirds which use it. In Colombia, the home of the most famous coffee farmer in the world, Juan Valdez, over 60 percent of the coffee exported is sun-grown.

Coffee is the second most valuable commodity traded legally in the world today, after oil, and some 70 percent of the world's coffee is produced in Latin America and the Caribbean. In many of those countries, it accounts for 40 percent of their exports, so it is a keystone of their economies. And yet, most of the profit is reaped by only a few multinational corporations. The farmers live on very low wages; some earn no more in a day than you pay for a latte.

North Americans consume about one third of all the coffee that is produced in the world. We Canadians buy about 130,000,000 kilograms a year. To look at this another way, as consumers, we wield a lot of power in this market. We may have the feeling that the large corporations are not listening to our questions about the social and environmental problems of sun-grown coffees, but you can bet they will listen to our wallets.

In an age when consumers are becoming more conscious of threatened wildlife and global poverty, they are beginning to demand that their coffee not only be of good quality, but be grown and harvested in a way that is environmentally and socially responsible. "Organic" coffees are grown without chemicals. "Fair trade" coffees are produced in such a way that all people involved in the process earn a decent income in good working conditions. "Shade-grown" or "bird-friendly" coffees enhance biodiversity, are more likely to be organic and tend to be grown more on the *fincas* owned by small farmers. But how do you know what you're buying?

Just as you look for the "dolphin-friendly" logo when you buy canned tuna, you can read the labels on the coffee you buy. If you start with whole-bean coffee, you can be certain you're buying coffee and not fillers like chicory. Many coffee producers are now seeking certification from independent organizations, and certification labels appear on the packaging. Ethically produced coffees are more widely available than you might expect. Several roasters in British Columbia are marketing them, and some non-profit organizations also sell coffee produced by cooperatives in Central America. Expensive? These coffees are very competitive with other high quality whole-bean coffees.

If you don't drink coffee, you probably eat chocolate. As an interesting aside, the story is much the same with cacao, from which chocolate is made. It is now possible to buy chocolate which is certified organic and fair-trade.

Every day we are assaulted with relentless reports of global poverty and hunger, of environmental devastation, of deaths and extinctions in the name of profit. It all seems too overwhelming for us to imagine a way to help. Well, we can't solve all the world's problems at once but we can take an important step by making better decisions as consumers, and by spending our dollars ethically.

Resolve now, before the year is out, that you will buy shade-grown, organic, fair-trade coffee. Ask for it in the grocery store, ask your barrista for it. Explain why you think it is important. Give it as gifts, to start your friends and family on the same path. By speaking out as consumers, we can have a profound impact on the lives of the coffee farmers, and the birds that share their fincas. You'll never enjoy a cup of coffee more.

January

Dark-eyed Juncos

The Rush Birds

I have filled a lot of bird feeders in my time, in a wide variety of backyard habitats, from retired agricultural fields to Garry Oak hilltops to Douglas-fir woodlands. The birds vary, of course, with the habitat but there are a few constants that seem happy almost anywhere.

Every morning, they are there: in snow and rain, in sunshine and stormy weather, in such numbers that it is impossible to know for sure how many there really are. These are the juncos, a mainstay at many feeding stations, from September through April.

Times are a little simpler now. Way back in the early 1970s, we had several juncos to choose from, but now they are all lumped together as one species that is called the Dark-eyed Junco. The Latin name for the species is a curious one: *hyemalis*, which is derived from the Latin and Greek meaning "winter." It was apparently suggested by the early ornithologist Mark Catesby, who only knew the species in winter, and in the southern part of Canada this winter visitor is known widely as the "snowbird." The genus or family name Junco, however, is from the Latin meaning "rush," and none of the references I've consulted can explain why this bird that loves coniferous woods and edges is forever associated with plants of the shoreline.

The junco we see here most often is the race known as the Oregon Junco.

The male sports a soft brown coat over a white belly, and has a striking black hood sharply cut off from the other colours. His bill is bright pink and his eyes, as his new name suggests, are dark. The Oregon Junco is a winter visitor to urban feeders on the south coast, but they also nest here not far back in the woods.

Occasionally in winter we are visited by a cousin from the east, the Slate-colored Junco, and the name is a fitting one. Its upperparts are all slate grey, with no distinct hood, but it retains the whitish underparts of its cousins. This is an elegant little bird at any time, but arriving amid a group of Oregon Juncos, it is quite striking. All juncos display white outer tail feathers when the tail is fanned or when the bird takes flight; it is a nice, unexpected finishing touch.

At this point, however, the junco story gets a little muddy. In Canada there are seven subspecies of Dark-eyed Junco and unfortunately, the Slate-colored and the Oregon forms are at the two extremes of plumage variation. In between, on Vancouver Island we also see a race much like the Oregon Junco, but a little paler, with a little more grey on the back and a hood that is less solidly black.

Add to this the fact that females tend to be less strongly marked than males and you have a lot of room for overlap. Finish off with a healthy tendency to hybridize and you have more than enough juncos to fill in all the gaps that ever existed and helped you to separate one subspecies from another.

So you will find at your feeders birds we might call For-The-Most-Part-Oregon Juncos, a blend of browns and pinks and salmon colours, with hoods ranging from pale grey to black. They seem unperturbed by differences in their cousins and travel together in large, loose flocks throughout the winter.

They spend much of their time foraging, feeding mostly on weed seeds; they will take millet and black sunflower readily at feeders. When they descend on a feeding tray, the air is filled with rapid-fire soft kissing noises as they call to one another and jockey for the right to a place at the bar.

As the days lengthen, the numbers thin out a little, and the males begin to test their voices for the breeding season to come. The song is a steady trill, not very musical, but not so dry as that of a Chipping Sparrow. And combined with the promise of spring in the air, it sends a little shiver through my bones each time I hear it.

With the coming of May, virtually all our juncos will be gone from urban areas. Some will move north to familiar territories, others just far enough into the woods to find a suitable nest site on the ground, or perhaps concealed in the roots of an upturned tree. They will raise probably four or five young, but in the harsh reality of their first year, few will survive. Banded birds that have lived to the age of eight or 10 years have been found, but this is not likely to be the norm.

Next fall, we'll see juncos returning again to our feeders. Many of this year's visitors will probably not survive the gauntlet of storms, house cats and Sharp-shinned Hawks, but we can be sure that some will make it. While we won't be able to recognize them by looking at them, their return will be evidence that they have been here before. They may even look as though they're as glad to be back as we are to see them.

By Any Other Name

What's in a name? Bird names are as rich in their diversity as the creatures they identify. The Red-tailed Hawk has a red tail. The Cooper's Hawk is named after a man named Cooper. Shaggy leg feathers have given the Rough-legged Hawk its name, but is the same true of the Rough-legged Buzzard? There are Pigeon Hawks and Common Nighthawks, but how are they related to one another?

The fact is that the nature of their relationships varies considerably. The Common Nighthawk is not a hawk at all. Pigeon Hawk is an old name for the Merlin, which is not technically a hawk, but a falcon. The Rough-legged Hawk is the same bird as the Rough-legged Buzzard — it just lives in a different part of the world. And even though the Cooper's and the Red-tail are shaped quite differently, they are very closely related.

The confusion arises in the use of the common names for these birds, which often give no indication of the birds' relationships to each other, and vary from culture to culture. To clarify the situation and avoid confusion, the scientific community uses its own standardized set of names for all living things. The names are Latin, and they are understood worldwide. They follow a system that is known as the Linnaean system, credited to the 18th-century Swedish botanist, Carolus Linnaeus.

Linnaeus was born Carl Linné, on May 23rd, 1707. (There is confusion even here, depending on whether your reference is the Georgian or the Gregorian calendar.) The son of a curate, Linné had no interest in following in his father's footsteps; he was fascinated with the natural world. Very early in his life, he made detailed descriptions of the plants he encountered.

Fortunately, his family was encouraged to send him to medical school, which allowed him to pursue his interest in natural history as well. Linné became convinced that a standardized system of naming plants was necessary and that Latin was the language that should be used. He was so enamoured of this idea that he Latinized his own name to Carolus Linnaeus, and that is how we know him.

Where do birds fit into the Linnaean scheme of things? Well, in the grossest of categories, they are members of the Kingdom Animalia, one of five kingdoms in the world of living things. This ties them to such diverse creatures as humans and slugs. The level below this is called the Phylum; birds belong to the Phylum Chordata, which includes all the vertebrate animals (the sub-phylum Vertebrata) and a few odds and ends. Birds become distinguished from other vertebrates like reptiles and fish and mammals at the next level of definition, called Class.

Birds comprise the Class Aves. Within the Class Aves, all living birds are placed in the Subclass Neornithes, the true birds. This separates them from the Subclass Sauriurae, which includes extinct toothed birds like Archaeopteryx, and the Subclass Odontoholcae, represented solely by the extinct loon-like Hesperornis.

The Subclass Neornithes is further broken down into three Superorders. The Palaeognithae includes only the fowl-like Tinamous of South America. The Ratitae includes the flightless land birds, like the Ostrich and the Emu. All other birds fall into the Superorder Neognithae.

Within this Superorder, there are many Orders. Some, like the Passeriformes, include thousands of birds like our warblers and sparrows. The Order Coliiformes, the mousebirds of Africa, has only six species.

The Red-tailed Hawk is in the Order Falconiformes, as are the Cooper's Hawk, and the Rough-leg and Merlin. This order includes all the diurnal raptors. The Common Nighthawk is actually a nightjar, in a different order called the Caprimulgiformes.

The Falconiformes break down into three or four Families, depending on the authority you use. The Family Cathartidae, including three species of North American vultures, once belonged here, but has recently been moved to the Order Ciconiiformes, to acknowledge the close relationship with the storks of that order. Secretary birds comprise the Sagittariidae. The Falconidae include all the falcons, of which the Merlin and Peregrine Falcon are examples. The Osprey is sometimes listed in its own order, the Pandionidae, and sometimes included with all the other hawks and eagles, in the Family Accipitridae.

Below the Family level, every bird has a unique scientific name. The first part is the genus and the second is the species name, both usually italicized. The Red-tailed Hawk is *Buteo jamaicensis*. There are other hawks with the name "*Buteo*," but none with "*jamaicensis*." Similarly, the Cooper's Hawk shares its genus, *Accipiter*, with other species, but no other uses the second name *cooperii*.

Finally, many species are broken into subspecies, which may or may not be distinguishable, but which could or do interbreed freely where they occur together. Thus the Western Red-Tailed Hawk is *Buteo jamaicensis calurus*. Its cousin, the Rough-legged Hawk, is *Buteo lagopus*. Linnaeus himself, in Sweden, probably saw the same species, but knew it as the Rough-legged Buzzard. Scientists call the subspecies found in Sweden *Buteo lagopus lagopus*, while the subspecies that occurs in North America is *Buteo lagopus sanctijohannis*.

The work of Linnaeus has been modified and improved over the years, but it remains the foundation of today's system of scientific nomenclature. In his own time Linnaeus was recognized for his achievement: he was knighted by the Swedish government, which allowed him to be addressed as Carl von Linné. He died on January 10th, 1778, and is buried in Uppsala Cathedral, an honour equivalent in England to being buried in Westminster Abbey. We know him as Carl Linné, Carl von Linné, and Carolus Linnaeus, but by whatever name, science would not be the same without him.

From the Sea to the Continental Divide

It was much like any other identification problem. All we had to do was get the field marks, and then hit the books.

"Raven-sized. Tail dark brown; feet and legs blue; upperparts, brown, shading darker on the belly; face and breast, yellow, with a red band across abdomen; bill half as long as body; lower mandible, chestnut; upper mandible, yellow terminally and chestnut basally." I had done enough reading to know what this bird was, but even so, I was left with my mouth open as it moved quietly through a 50-metre almond tree. The book confirmed that Chestnut-mandibled Toucans are common here, in the Caribbean lowlands of Costa Rica.

A gang of Collared Aracaris moved in next, smaller than the toucans, but just as gaudy. They picked up tiny berries with their enormous, banana-shaped bills, and deftly tossed them back into their throats. Orange-billed Sparrows skulked through the understorey, looking like ornaments in a tropical-plant store. The air was sultry, with no breeze; one of the frequent rain showers had just ended. Tiny poison-dart frogs, bright red in warning, came out from under leaves and logs. An unidentified denizen of the jungle flew to the cover of a raffia palm, with leaves the size of umbrellas.

After lunch at Tortuga Lodge, the guide shepherded our group of visitors into a flat-bottomed boat, and we headed off into the network of canals that parallels the coast, and provides the only transportation route through the rainforest.

Upstream, crocodile-like caymans slid into the water, and a Jésus Christo lizard lived up to his name, running broken-field style across the surface of the water. Around a bend, we lost about a million years, as metre-long iguanas moved along branches or swam to safety. Limpkins, heron-like but more closely related to the rails, couldn't decide if they were afraid or curious.

Costa Rica is a country that has become famous for its natural history. Roughly the area of Vancouver Island, it has as many species of birds as there are in all of North America. The land rises from sea level to almost 4,000 metres, and there are over 80 volcanoes in the country.

The climate is not all tropical, even though it is only 10 degrees from the equator. One of the most celebrated ecosystems in the Americas is found at Monteverde, a biological reserve that straddles the Continental Divide. Here, sweaters are often in order, and rain gear and boots are a must. Even in the "dry season," wind-driven mist constantly enshrouds the mountaintops; it truly lives up to its name: cloud-forest.

In this forest, there is a diverse cast of characters. The tall trees are burdened with communities of epiphytes, plants that secure themselves to the trees but get all their moisture from the air. Prehistoric-looking tree ferns grow up from the forest floor.

The bird life is hard to see. The dense foliage conceals most birds, but some are a little bolder. Hummingbirds come out of hiding to investigate a bright red rain jacket. Now we realize that almost all the trees and shrubs blooming in the forest are bright red. Hummingbirds are confined to the New World. Most British Columbians will see one or two species, and occasionally a vagrant will appear briefly. In Costa Rica, there are 50 species. These dazzling little birds seem out of place in the darkness and rain of the cloud-forest, but they are brassy and territorial. Their colours are spectacular, especially when the sun catches the iridescent gorgets on their throats. Their names are evocative: Magenta-throated Woodstar, Violet Sabrewing, Green-crowned Brilliant.

But the bird which probably brings more people to Monteverde than any other is different. It is up to a metre in length, counting its tail plumes, and is clad in emerald greens, with a scarlet abdomen and a white undertail. It does not surprise us that the natives used to worship the bird they called quetzal. And now birders make the pilgrimage to see — no, to worship — the Resplendent Quetzal.

When I travel to a country to see something special, I am wary of being disappointed, and I often reserve hope of seeing it. But on the day I was at Monteverde, the quetzals arrived, right on time and just above the sign that says "Quetzal." Visitors drank in the spectacle for over an hour.

The walk back to the pension where we were staying took us past farm fields, with dairy cattle owned by the local Quaker community. Slate-throated Redstarts moved through the roadside greenery, and a Mountain Robin sang from a tree-sized poinsettia.

The wind blows hard at Monteverde and as we wound down the dirt road the next day, we could see the clouds blowing over the tops of the mountains. Relieved of their moisture here, they would move harmlessly to Costa Rica's Pacific Coast — the dry country. We followed them out of the mountains.

Tambor

The Pacific slope of Costa Rica is a smorgasbord of habitats. In the south, tropical forests grow lush and green, and the fauna are much the same as on the Caribbean slope. At Carara, about halfway up the coast, there is a definite transition to the drier and more open savannah type of habitat, and patches of dry tropical forest.

The Biological Reserve at Carara stands out from the air as the last patch of forest in the area. From the ground, within the reserve, the visitor can appreciate what has been saved in this small tract.

Troops of white-faced capuchin monkeys greeted the group, throwing debris and feces down on us, as they do on all intruders. Their agility in the treetops left me open-mouthed in awe, but for them, it's merely a part of their daily routine. Several species of trogons, brightly coloured but resting unobtrusively overhead, tested our knowledge of the field marks. The endemic Baird's Trogon was an extra treat.

We were aided at Carara by Rodolpho Zamora, a passionate young guide who did not have enough hours in his tour to show us his Costa Rica. His eyes, accustomed to the tropical forests, found for us a Rufous-tailed Jacamar, and giant blue Morpho butterflies, refulgent in the shadowy canopy of the trees. He showed us the nest cavity of a pair of Scarlet Macaws, endangered now, but safe, at least at Carara. There in the opening, both birds appeared, like a postcard, but moving, alive! On our walk back, Rudy brought us close enough to a group of peccaries to hear the anxious rattling of their tusks.

North of Carara, the Pacific slope receives much less rain. January is the height of the dry season as well, so it was hot. In Costa Rica, the dry months are thought of as summer, even though they occur during the season we normally think of as winter. Many *ticos* — Costa Rican locals — take their holidays at this time, and the splendid Pacific beaches can be busy.

The Nicoya Peninsula, in Costa Rica's northwest quarter, is very different from the rest of the country. You can explore the area by road from the north, or take one of two ferries from the decadent fishing village of Puntarenas, past its prime in an era of increasingly industrial fishing operations.

At the ferry dock, I looked up to see large numbers of huge birds soaring. Black Vultures and Brown Pelicans hung in the breeze, sharing the skies with Magnificent

Frigatebirds of all ages and sexes, each looking like a different species. I wondered if a southern visitor to a Vancouver Island ferry terminal would be as impressed by the Glaucous-winged Gulls, Bald Eagles and Northwestern Crows.

The foot ferry docked at Paquera, and from there it was a rough, 50-minute trip by four-wheel-drive taxi to Tambor. We had come to stay at a working cattle ranch; the hacienda had been expanded and augmented to accommodate guests comfortably in a hotel-style setting. The beach is several kilometres in length, and in a week's stay there, I did not see more than a dozen people on it at any one time. There were more shorebirds than humans: Willets, Whimbrel, Black-bellied Plover and lots of Spotted Sandpipers (unspotted at this juncture).

Around the hacienda, there were gaudy White-throated Magpie-Jays, and White-crowned Parrots. In a tidal pond, a pair of Green Kingfishers courted ebulliently. From an excellent vantage point on a covered patio, I could see Roseate Spoonbills, Cattle and Great Egrets, Green, Little Blue and Tricoloured Herons. A Bare-throated Tiger-Heron alternately lurked and paraded. White Ibises and Wood Storks flew overhead. The Holiday Inn was never like this!

No matter where you are, the best time to bird is at sunup and at Tambor, most days I was out a couple of times in the morning, avoiding the afternoon heat by sitting at a shady poolside and catching up on notes. The breeding activity had not yet started in earnest, and many birds were quiet. But in the tropics, birds often move in mixed flocks, and on a couple of occasions I found active groups that both excited me, and challenged my skills as a birder.

In a shady creek bottom, a Turquoise-browed Motmot shared a tree with a pair of Black-headed Trogons and a Squirrel Cuckoo. A nearby tree laden with fruit held Scissor-tailed and Streaked Flycatchers, and the ubiquitous Tropical Kingbirds and Hoffman's Woodpeckers. Tennessee Warblers were everywhere, and Baltimore Orioles and Hepatic Tanagers also added a North American flavour at times.

These northern species are really at home here, and merely visit Canada for a few months in the summer. Their existence depends on the foresight of people like the Costa Ricans, and I was quietly grateful for the way in which the ranch operations accommodated the northern visitors, birds and birders alike.

When we boarded a small plane to fly back to the capital city of San José,

I watched as a small knot of men moved across the makeshift grassy airfield. Their shirts were sparkling white in the sun, their pants neatly pressed, with narrow cuffs breaking perfectly over fine leather loafers. Hands unrolled large sheaves of paper, and fingers pointed in various directions. The hacienda's new owners, Spaniards, it was said, were here, discussing the location of a new beach hotel. I knew as I left that I would probably not be returning soon but this time, perhaps, it is for the best.

Anna and the Duke of Rivoli

It seems an unlikely thing in January to be thinking about hummingbirds. Well, sure, maybe you just got back from Costa Rica and you just know your friends will want to hear about the Violet Sabrewings you saw, while they were scraping the frost off their windshields so they could drive to work in the dark and then drive home from work in the dark, too. Truth is, your friends might have been watching hummingbirds too, right here in coastal British Columbia, in the middle of winter.

The hummingbird we see on Vancouver Island in winter is called Anna's Hummingbird. It is a fairly recent arrival this far north, but because of several factors, it is almost certain to remain a part of our avifauna.

Anna's Hummingbirds historically ranged from northern Baja California to northern California. The species is sedentary; that is, it does not migrate as do most other North American hummers. But they do wander, and they have expanded their range northward along the coast, with the first vagrants reaching Victoria in the 1940s. Their status changed from fluky to rare through the 1960s and, since then, their numbers have grown steadily. They now occur north as far as Nanaimo and on the Lower Mainland as well. First drawn to jasmine and other winter-flowering plants, the Anna's are now most often reported at feeders. Much of their diet in winter, though, is spiders, so the healthy gardens of our mild coastal climate are a draw in themselves.

Although the Rufous Hummingbird is by far the most common of the four species that occur in B.C., any hummer seen on the south coast in the winter months is almost unequivocally an Anna's. While the Anna's are certainly less numerous overall, a migratory Rufous Hummingbird here in the winter would be extremely rare indeed.

For many years Anna's were uncommon and very local. Once they find a good bit of habitat with a steady supply of food, they aren't inclined to wander, especially in the winter. If the steady supply of food happens to be a hummingbird feeder in a sheltered garden, then that's probably where an Anna's will set up shop.

For years, if birders wanted to see an Anna's Hummingbird, they asked around to see who had one coming to a feeder. Now, their numbers are increasing to the point that they are being reported commonly over much of southern Vancouver Island. While there are records from a surprisingly large area of the province, it is only the south coastal area where they are able to survive the winters.

How do they survive? In mild winters, they can subsist on the nectar from winter-flowering plants, and insect matter, but a prolonged hard freeze would probably kill them off. Enter the hummingbird feeder. With a steady supply of carbohydrate, the hummers are able to survive surprisingly bitter weather. This is in large part due to the dedication and ingenuity of those feeding station operators who have kept their feeders flowing, even when the weather has turned frigid and snowy.

Anna's, like most hummingbirds, also have another trick up their sleeves. Overnight, they slow their metabolism dramatically, so that less energy is consumed. This way, they can survive a long winter's night without feeding.

Most field guides should help you to identify Anna's Hummingbirds pretty readily. The males of this species are iridescent green on the back, greyish-white underneath, and have an iridescent fuchsia-pink gorget that extends from under the chin up and onto the forehead. The females are similar, but they have no gorget, except for an occasional few dots or a small spot in some birds. Male Rufous Hummingbirds, by comparison, are mostly cinnamon in the upperparts. Females are similar to Anna's, but a little smaller, and always show some buff or cinnamon on the flanks, and cinnamon in the spread tail.

With the occasional appearance of a Xantus's Hummingbird or a Costa's Hummingbird, both vagrants from farther south, at B.C. hummingbird feeders, people begin to see weird and wonderful things at their own feeders. Just remember the odds — it's extremely unlikely that another species might accidentally wander to your feeder, of all places.

And who was this lady named Anna anyway? Anna was the wife of Victor

Massena, the Duke of Rivoli and Prince of Essling in Italy, who was a keen amateur naturalist. A specimen brought to the duke by collector Paolo Botta was described for science by his friend René Lesson, who named the bird after the duke's wife, to honour her shared interest in birds. The Massena collection of 12,500 specimens is now held at the Academy of Natural Sciences in Philadelphia.

If you discover that you have Anna's Hummingbirds in your garden, keep feeding them for as long as they continue to come. In the spring, when the Rufous return, there'll be a little jockeying for position, but they seem to work it out. If you'd like to reduce the skirmishes, try putting another feeder out some distance away, and out of sight of the first if possible.

There are now several breeding records for Anna's Hummingbirds from southern Vancouver Island. They begin to nest very early and many of the nests have been recorded in February, so it is a good time to start watching for courtship behaviour. Look for the males doing their carnival-ride courtship display flights. And the females may be seen collecting bits of spider web, which they use to bind their nests.

In recent years we may actually be seeing an increase in numbers as a result of a run of mild winters. But this increase is also a result of the artificial availability of food in the months in which a pioneering Anna's might have perished. Just as humans and their bird feeders have aided the expansion of the Northern Cardinal's range in the east, we have had an effect on the range of the Anna's Hummingbird. Unnatural? Perhaps. Wrong? That's a tough one to answer, but I don't think we've upset the balance of nature too much on this one, so let's just enjoy them for as long as they choose to call our gardens home.

Sharp-shinned Hawk

February

A World I Know a Little Better Now

Just before Christmas one year, my wife Jan presented me with an early Christmas gift. It was plainly wrapped, in an old yogurt container, although her sense of occasion had prompted her to put a bow on it. It was a gift, she said among a group of friends, of which only I would appreciate the significance.

Inside the gift container was a collection of moist grey lumps, with bits of wood chips and grass clinging to them. A smile crossed my face. I have seen many owl pellets in my time, and I knew right away what I was looking at. Pellets are the indigestible remains of an owl's meals, the fur and feathers and bones of its prey. They are cast up some hours after a meal, and their discovery means one thing: where the pellets are, an owl has been. I knew from the debris clinging to them that these had come from the paddock outside our little barn. So the significance of these first pellets was their attestation that a Barn Owl was visiting.

Four years previously, we had built the barn to accommodate a few of the usual barn animals. But we also incorporated a nesting box that we hoped would attract Barn Owls. We knew that the owls nested in a derelict silo nearby, and we reasoned that at some point the offspring of one generation or another would need to seek out new territory.

We took advantage of an opportunity to increase our chances the following summer. We arranged for six young Barn Owls, rehabilitated at the Orphaned Wild Life (OWL) facility in Delta, British Columbia, to be released from our barn. Faced with dwindling Barn Owl habitat and nest sites in the Fraser Valley, the centre was looking farther afield for suitable release sites for its birds.

In June, two workers arrived from the centre, the six young owls having experienced their first ride on a B.C. Ferry. Three males and three females, they all looked to be in fine shape, full-grown and in beautiful Barn Owl plumage.

After the requisite photo opportunity, the birds were released. One by one, they flopped out of the open hayloft door and took in their new surroundings. The crows found them immediately, but the owls seemed unflustered by the harassment. They disappeared quite quickly in various directions, most flying into the cover of tall trees.

The success rate of these releases is disappointingly low, but it is no worse than the odds for the survival of Barn Owls in the wild. Within a year, it is likely that two of these birds would die of unknown causes, one would be killed by a car, one would fall prey to a Great Horned Owl, and one would starve. One might survive. After the first two days, we did not see any of these owls again. Owls continued to raise young successfully nearby, but we had no way of knowing for certain whether any of the released owls were among the nesting birds.

Our nestbox remained empty. Oh, a few Rock Pigeons ran away from their dovecote up the road, and roosted in the nestbox for a while, but there were no signs of owls in it. During one severe winter storm, our neighbour's efforts to keep his silo standing for the owls were defeated, and the wind brought it to the ground. The resident pair of owls must have renested elsewhere, but still there was no activity in our barn.

So the discovery of the pellets the next winter, just below the opening to the nestbox, renewed our hopes that a pair of owls might take up residence. The following evening, we discovered more encouraging evidence: three more quite fresh pellets inside the nestbox. We became immersed in the lives of these owls. We listened to their screams and bill-clicking as they established their territory. We watched as they flopped silently overhead, ghostly in the beams of our flashlights. And we held our breath as they returned to the barn, again and yet again.

We checked the box in darkness, so that if an owl was disturbed inside, it could fly into the familiarity of the night. When we discovered the first egg on one visit, it was a cause for great celebration.

That year, the owls raised seven young. At regular intervals, we made short visits to the nest to monitor their progress. It did not appear to affect the nesting, but still we were a little uncomfortable about our intrusions into their lives. We were anxious to learn more about the birds, but afraid that we might discourage them forever.

After the first year, the owls returned to nest every winter. We recognized some of the birds, and realized that sometimes one member of the pair was different. One easily identifiable female was a rich cinnamon colour underneath, but she was present for only one nesting. Some years, fewer eggs were laid, and sometimes one or more of the young died in the nest. One female (we think a young bird) laid two very small eggs in an unused corner of the box. She never incubated them, but laid a normal full clutch shortly afterward.

Andy Stewart, a raptor biologist, offered to band the young birds, to contribute to research already being done on Barn Owls in B.C. Using a long ladder, we retrieved the birds, all down feathers and talons, one by one and brought them into the tack room — now a makeshift banding station. The defiant screeching of these young birds is other-worldly. So is the smell of the Barn Owl excrement that unfailingly finds its mark on your clothing. Once back in the nestbox, the banded young settled quickly, and the female returned within a few minutes.

A few years later, I was presented with another small and curious parcel. Bob Chappell, a keen birder in his retirement, handed me a little black object, smaller than a deck of cards. I looked as politely impressed as I could. "It's a video camera," Bob said. "I want to put it in your Barn Owl box."

Bob installed the miniature video camera and a microphone in the nestbox. A television monitor and several VCRs were installed in the tack room, now renamed Mission Control. Eight hundred kilometres of assorted electrical cable festooned a barn that heretofore had seen nothing more technologically advanced than a manure fork.

The owls accepted the arrangement with equanimity, and during the daylight hours we watched as the female incubated. She was the picture of patience, sometimes allowing herself to nod off. Occasionally she would readjust the clutch, and it is a testament to the engineering integrity of eggs that her sometimes rough handling did no damage.

Barn Owls are strictly nocturnal, and some of the best views we had came toward dusk, when the young became more active. Occasionally we would see the arrival of the male, with food, but the light faded quickly and the videotaping stopped for the night. We experimented with low-level artificial light, and with light from a red bulb, but this seemed to bother the adults, so we gave up on it. Bob went to work on a solution.

He came up with a small circuit board that held about a dozen light-emitting diodes which produced infrared light — visible to the camera, but not to humans and not to the Barn Owls. It was an immediate success. We could now film during the black of night, when the birds were most active. For the next several years, Bob put in many weeks refining the system, all the while filming hundreds of hours of Barn Owl activity.

Much of what we saw confirmed what we had read about these beautiful owls. Barn Owls incubate their eggs as soon as the first is laid, so the first one has already developed for two days by the time the next is laid. With the rest laid at two-day intervals as well, there is a spread of two weeks or more in the ages of the youngest and oldest.

The young and the female are fed by the male for some time. When the young are able to thermoregulate on their own, the female leaves the nest to assist in the hunting. If there is a good supply of prey, and if both parents are good hunters, all the young are likely to survive. If there is a food shortage, the smallest of the young will not receive enough to eat and will die.

The video footage revealed some things that do not appear to have been documented in the literature. On one piece of tape, Bob discovered what is apparently a third adult bringing food to the young. The response of the young owls to this bird was entirely different than to their parents. They shrieked defensively and backed into the corners until the mystery adult left, whereupon they wolfed down the prey.

The young birds fledge when they are about five weeks old, and begin to follow the adults as they hunt. The adults must still return to the nest to feed the young that have not yet fledged. We would eventually hear the whole family of young owls nearby, begging for food. The raspy hiss is incessant, at least until their all-day hunger is satisfied.

Early in the fledging period, the adults begin to be more vocal too, with their chirrupping and castanet-like bill clicking. They continue to be very territorial. Outside our bedroom window the owls would sometimes perch on two knee braces which support the end of the roof. It made a charming picture, save for the adrenalin surge when we were awakened, hearts pounding, by the chilling screams of the male claiming his dominion.

Over the years, we became more than a little attached to the Barn Owls. Although their calls at night became as familiar to us as a robin singing in the garden, I never tired of watching them, coursing the night sky and just being Barn Owls.

My own family has fledged and we no longer share our lives with the Barn Owls. I've moved away from the agricultural habitat that is home to these birds and now live surrounded by Douglas-firs. My home office looks across a mossy outcrop and, on the other side, affixed to a tree where I can see it clearly, is a large wooden box. It has a hole in it, just the right size for a Screech-Owl. I have my fingers crossed.

My bookcase looks like a bookcase in many other homes, with photographs of my three sons along the shelf. But there is also a family portrait of seven very unkempt youngsters, with fluffy down surrounding their faces, and black eyes peering out at me, from a world that I know a little better now.

This Business of Flight

A little over three centuries ago, Daniel Bernoulli was born of Swiss parents in Groningen, in the Netherlands. Bernoulli was not an ornithologist, or even a birder as far as I know, but his work was as critical to our understanding of birds as that of any collector or taxonomist or geneticist.

Bernoulli was a mathematician and physicist who is best known for Bernoulli's Principle, which (for our purposes) states that the faster a fluid moves, the less pressure it exerts. What Bernoulli didn't know is that his principle would lead to our understanding of how flight is achieved. Birds, of course, having the undisputed ability to fly, have had better things to do than to wonder how it happens. But it continues to amaze us land-bound bipeds, so we owe a small debt of gratitude to Signor Bernoulli.

In a wing, or airfoil as it sometimes known, the stream of air which meets the leading edge as it moves through the air is split: part flows over the upper surface, and the balance flows beneath the under surface. If the upper surface is convex and the lower is flat or concave, the air passing over the upper surface must travel farther than the air passing beneath the wing before they meet at the trailing edge. Because it must travel farther, it travels faster.

Here is where Bernoulli comes in. Because the upper airflow is moving faster than the lower airflow, there is less pressure on the upper surface of the wing than on the lower. This lower pressure acts as a sort of vacuum, "sucking" the wing up. That is what gives birds the lift they require to stay airborne. And the same principle is what keeps your jumbo jet and the ground a safe distance apart.

So far, so good. If you are a gannet or a puffin, and you normally get airborne by jumping off a cliff, you can manage to stay up. But what about all those birds that have to get themselves up in the air without the aid of cliffs?

This is where flapping flight comes into play. A bird's wing is divided into two basic parts, the inner and the outer wing, or "hand." The inner wing provides most of the lift, and the outer wing provides the forward motion. The feathers of the outer wing are very flexible and manoeuvrable, and they act as miniature airfoils in themselves on each stroke of the wing.

On the downstroke, the wing moves forward and down, and the primary flight feathers are canted so that each one produces lift. The bird controls the angle of the wing and the feathers, so as to gain the right combination of vertical and horizontal momentum.

On the upstroke, the flight feathers of the outer wing perform no function, and are turned edge-on, or "feathered," to reduce resistance. The inner wing is tucked close to the body to further streamline the wing. When the need for power has passed, the wing can be spread for gliding flight. The forward momentum causes air to move over the airfoil shape of the wing and provides the lift required to keep the bird aloft.

In still air, a bird in gliding flight will eventually run into a problem.

Because of the drag caused by the resistance of the bird's body as it passes through the air, it slows down. As it slows, the lift is reduced, the bird yields to gravity and gradually returns to Earth. When a little more power is applied, speed and elevation are gained and the situation is again under the control of the bird. Even the puffins that can get airborne by stepping off a cliff must sometimes resort to flapping flight to maintain height.

At some point, though, our bird is going to have to lose elevation and speed in order to land without damaging its delicate airframe. This is accomplished in a variety of ways.

Loons rely on water landings to lessen the impact. Herons have large wings that provide a lot of lift (with a cost of reduced speed) and can let themselves down quite gently. But most birds rely on principles of flight in some way to achieve a trouble-free landing. Lift is increased and speed is decreased by increasing the angle at which the wing meets the oncoming airstream, both necessary factors in a safe landing.

If the angle is increased too much, then the air flow over the wing becomes turbulent, all lift is lost and the bird drops. To counteract this, birds have a small extra wing called the "*alula*," corresponding to the human thumb, which acts to direct air flow more smoothly over the wing, achieving yet greater lift as the bird's forward motion is reduced to the point at which a landing can be made. Here again, you will see the principle at work when your Airbus takes off or lands, as slots in the leading edge of the wing are deployed.

With getting up and getting down more or less under control, we can relax our seatbelts and take a closer look at some of the finer points of bird flight. First of all, we have seen how the shape of a bird's wing enables it to keep from tumbling to the ground. Once the bird is moving through the air, the forward motion causes sufficient air to pass over the wing to provide the lift that keeps the bird aloft. This simple flight is called "gliding."

All birds glide to some extent (though the jury is still out on hummingbirds) but they all must overcome the inevitable pull of gravity. We've all seen birds use flapping flight to maintain height and speed. Others do almost the opposite, and flap very seldom. How do they keep from gradually coming back down?

The answer is that they use their ability to glide in an active rather than a passive way, in the form of flight that is known as soaring. These masters of the air are familiar to us: the eagles and hawks, the gulls, and the albatrosses of the open oceans.

In order to stay aloft for long periods, most of these birds take advantage of air which is "going their way" — up. This happens under many circumstances. One example is what is known as an obstruction current. Air in motion arriving at a hill or building or ship is deflected upward; the birds know this and will ride the rising current to gain a little free height.

Perhaps the best-known aid to soaring birds is the "thermal," a bubble of warm air which rises because that is what warm air does. Thermals occur wherever there is a temperature differential, but are most common over exposed areas of open land, especially sun-warmed bare rock. In places like these, hawks, eagles and vultures rise with the thermals to great heights, sometimes with nary a wingbeat.

To take advantage of these sometimes unpredictable currents, hawks have evolved wings and tails which can be finely adjusted to make the most of the available updrafts. In contrast, albatrosses have evolved a different sort of wing, with which they engage in a different sort of soaring.

Over the open oceans, there are few reliable thermals; what is reliable is almost constant wind. This wind is slower nearer the water, because of friction as it drags along the surface. An albatross will glide with the wind, gaining speed as it descends, and then turn suddenly into the wind. Its momentum, meeting the oncoming wind, gives lift to the bird. As the bird rises, it meets air moving more quickly, and even greater lift is created. When the albatross's upward momentum dies, the bird turns to begin another glide, and the cycle is repeated. This form of flight is known as "dynamic soaring."

We cannot leave a discussion about bird flight without paying homage to the stunt flyers. These are the guys that seem to break the rules and still stay airborne. Foremost among these are the birds that have the ability to hover.

Many birds have the ability to stay in one place while in flight, but most of them require some assistance to do this. For example, large birds such as Ospreys appear to hover while actually flying very slowly into a wind. If the wind is strong enough, no flapping is required, and this is called "kiting." Other birds, like American Kestrels, are able, for short periods of time, to flap their wings so that their powered lift is equal to the gravitational pull exerted on their bodies. Some, with a little help from the wind, can even back up a little.

However, only one has complete freedom to fly in any direction it wants, or to stay put, at will. That, of course, is the hummingbird, and its body is as unique as its capabilities. The hummingbird, unlike other birds, flaps its entire wing in a figure-eight pattern. Each wingbeat can provide lift in either forward or reverse; all that the hummer needs to do is control it. And all of this happens 60 times per second.

For the birds, this is all pretty mundane stuff. For landbound observers like us, it has captured our imaginations for centuries. While our knowledge has come a long way since Bernoulli's time, we can thank him for shedding some of the earliest light in our search for a better understanding of birds.

The Unusual Among the Usual

I'm often not a very gracious guest, because I find myself distracted by the birds that are coming to my host's feeder. Fortunately, people are pretty tolerant of my transgressions, so I can get away with it when I'm visiting.

Among all the siskins and finches, I noticed at a friend's feeder the other day a Dark-eyed Junco which had a lot of white splashed across its cheeks. There was no question that it was a junco; its shape, behaviour and call notes were good clues. The white in its plumage must be the result of albinism.

This isn't always the case: Sometimes a bird that looks for all the world like something else will show up at a feeder. Usually, though, it is just one of the more common winter visitors, with some evidence of albinism which has altered the normal pattern of its plumage.

The incidence of albinism is quite rare. It occurs, to one degree or another, in only about .05 percent of all birds. That's about one bird in every 2,000. It can take many forms. In extreme cases, there is a complete lack of pigment. All the plumage is white, and all the "soft parts" (the feet, eyes and bill) are pink. But this occurs in only about 7 percent of birds exhibiting any degree of albinism.

Sometimes the pigment is absent only from the plumage or the eyes or the feet. Thus, a Northwestern Crow may have white plumage, but still have black eyes and feet. This is known as "incomplete albinism." Birds that exhibit abnormal patches of white, like the junco I saw, are more common. These are called "partial albinos."

The same effect may be the result of other causes, too. An injury can cause white feathers to grow in at the site of the wound. There are also records of birds that have been frightened badly, growing patches of white feathers after their next moult. At times it's a result of hybridization somewhere in the lineage.

Another condition closely related to albinism is called "leucism." It manifests itself as an overall paleness in the plumage. A similar condition is called "schizochroism," in which only one of the normally occurring pigments is missing. Either condition can produce birds that appear to be pale or buff instead of darker brown. These birds are often mistaken for Snow Buntings in basic, or winter plumage.

An anomaly that causes the opposite effect is called "melanism." In this case, there is an excess of pigment in the feathers and the result is a bird that looks much darker overall. This condition is apparently unrelated to albinism, and it is thought to be much rarer as well. One form of melanism seen fairly regularly is the so-called "dark morphs" of some of our hawks.

There is evidence that some "albinism" is nothing more than the result of aging in birds. As birds age, they sometimes grow patches of white feathers, or show overall paleness in their plumage. Black-footed Albatrosses get much whiter as they age. (Bird writers are also known to acquire a distinctly silvery cast to their plumage.)

Why don't we see more albinism in birds? One reason could be that the vast majority of birds do not survive into their dotage; the life expectancy of most small birds is one or two years.

Albinos, as we have seen, are quite rare. They also suffer at the hands of both friends and enemies. There is considerable evidence that albinos are shunned or tormented by others of their own species. That makes them more vulnerable, because they do not enjoy the safety of the flock. It is also known that predators will single out the odd bird in a flock; it is easier to spot it and follow it through to the conclusion of the chase. Simply put, albinos stick out. The net result is that albinos, scarce to begin with, are rare indeed if they have survived the attacks of their own kind and a host of predators.

If you're lucky enough to have found an albino at your feeder, it is probably rarer than, say, a Snow Bunting. If you are still unsure about its identification, compare its size, shape and behaviour with the other birds at the feeder. Does it have the pink bill of a junco? Is it the same size as the Golden-crowned Sparrows? Does it scratch backwards in the leaf litter like a Fox Sparrow?

The identification is a bit of a challenge but usually albinos can be sorted out pretty readily. The bad news is, they usually aren't around long enough for us to get to know them.

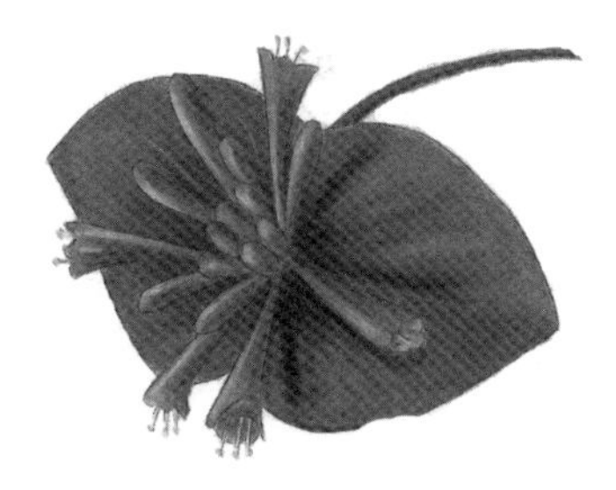

A Page from a Birder's Diary

December 1st

It came, as usual, from nowhere. Well, it came out of the northeast, I suppose. But when I looked in that direction a moment ago, it was not there. Then, quite suddenly, it was there, in front of me, well, out of nowhere.

Gyrfalcons seem to do this. Around Victoria, I'm accustomed to watching Peregrines racing across the winter skies, and sometimes stooping on their prey from great height. But the Gyr, the biggest falcon in North America, is different. It often hunts low over open ground, using hedges and other natural features to hide its progress. It is, sex for sex, larger and heavier than the Peregrine, and is built differently. It has broad wings for a falcon, but with the traditional pointed tips. The tail is wider, too, especially at the base, and this gives the bird a chunkier look. In flight, it is powerful and deceptively fast; it reminds me of a Pomarine Jaeger. Both are birds that shift from lazy flight to hot pursuit in the blink of an eye. With short, choppy wingbeats, the Gyrfalcon eats up the open country it prefers to hunt in.

Unlike the Peregrine, the Gyr's prey includes many mammal species, particularly on the tundra. It also preys heavily on ptarmigan. These ground-hugging species suit the Gyrfalcon's low-level pursuit flights.

But the big falcon also takes waterfowl, and that may very well keep this Gyr in the Martindale Valley north of Victoria through the whole winter. With a ready supply of Mallards and American Wigeon, a bird can feed itself through the winter and return to the breeding ground in good health, strong enough to defend a territory and feed a family.

I've been glassing the valley, looking for the Gyrfalcon, which has been here for a couple of weeks. There are lots of ducks to watch, and a couple of dozen Trumpeter Swans, honking tentatively from time to time.

A Glaucous-winged Gull gives a call that makes me turn to look at it. No eagle overhead to be alarmed about; no big raptor in the cottonwood tree; maybe just gull talk. No, there is a difference: now the Gyr is in my field of view, pumping low over the flats. As it approaches, it banks, circles and moves off to a power pole several hundred metres away. I watch it there for a moment; time enough to put the scope on it for a better look. The big falcon opens its wings, drops off the pole and powers low across the field again. Straight toward me it comes, looming larger. The focus wheel spins as the distance between us shrinks. Finally it sets its wings, drops its feathered legs and puts down on the mud. It is about 25 metres away.

This bird is an immature of the so-called grey phase of the species. It is streaked with brownish beneath and warm mid-brown above. Each feather on the upperparts is finely edged with white, giving the bird a scaly appearance. The eye, as in all our falcons, is dark brown. The unfeathered eyering, and the cere, at the base of the bill, are pale slaty-blue.

Turning from my notes to the scope, I can see what must have brought the Gyr back: a partly eaten male Mallard on the mud behind it. But the bird is not particularly interested in food just now. Instead, it hops about 10 metres to the left, where it stops, a couple of wing-lengths away from a female Northern Harrier. The two stand together nonchalantly for some minutes.

At this point, a gull flies in, perceiving no threat from a Gyr on the ground, and lands at the kill, where it picks away for several minutes. The harrier now puts up and sails away on those floppy wings. But it circles and returns, landing beside the kill and sending the gull back a few metres. The harrier then takes its turn at what can only be the Gyrfalcon's kill.

The scene plays itself out for several more minutes, with virtually no interaction among the three large birds. At some unknown cue, the Gyrfalcon puts up, circles once to gain a little height, and lands in the nearby cottonwood. There it becomes a poorly contrasted grey shape against a December sky, a falcon by shape, probably, and maybe a Gyr by its size and, well, it looks pretty plain, but gee, the light is poor.

You really need to see a Gyrfalcon fly to be sure you have seen one. Yes, you can identify them by their plumage and size, but particularly with immatures, they can be confused with Peregrines, and even sometimes with the distantly related Northern Goshawk.

But when that grey shape gets airborne, it moves so easily and quickly that you daren't take your glass off it for fear of losing it; they seem to come out of nowhere, and if you blink, they are nowhere to be seen.

Spring

Ruby-crowned Kinglet

Western Meadowlark

March

Beginnings

People make their start in birding in many ways. Some "have just always been interested." Others can pinpoint the day they became birders. For me, the precise date is gone, but some of the details are fresh in my memory.

I was a boy of nine when I went birdwatching that day with Mike. I often did things with my older brother, but when I look back, that outing was unusual because, at the time, I was not particularly interested in birds. I spent much of my spare time building plastic models of aircraft carriers.

I remember the day quite clearly — a little overcast and very warm. We were in a small bit of vacant scrubby pasture near our house in San Antonio, Texas. Even in the American south, the tree we approached was leafless as it contemplated spring. And I remember well that Mike was very anxious for me to see a bird. He held the binoculars in the right direction for me, pointed with his finger, and there on a branch, in unlikely technicolour, was a male Painted Bunting. It was, for a Canadian boy living temporarily in the American south, an undeniably exotic-looking creature.

My brother's interest in birds has continued to this day. I would like to say that from that point on, I too was hooked, but the truth is that I stepped up my aircraft carrier production.

In time, though, the fire for birds grew in me. When I joined my first Victoria Natural History Society field trip, the embers burst into flames and there was no looking back. Over the years, my passion somewhat overwhelmed my brother's more casual interest, and I find that now it is I who am anxious to show birds to him.

Although not often these days, we are occasionally able to spend a little time together in the field. Sometimes it is a walk in the Gatineau Hills near his home in Québec; now and again he passes through my neck of the woods and we take in some west-coast birds.

Today, we stop the car along Wallace Drive east of Brentwood Bay, opposite the little footbridge that leads to Willoway Trail. Our arrival is noted by an adult Bald Eagle, on a stout branch halfway up an elderly cottonwood near the beginning of the trail. As we get ourselves organized, the eagle pushes off and flies lazily over to another perch on the other side of the field. Beneath the tree, we discover the fresh remains of a Mallard, and not much of it at that.

Spring is definitely in the air. To our left, ducks quack and Trumpeter Swans are honking affably to one another. Pacific Treefrogs are tuning up for the mating season and a Marsh Wren sings a tentative first few notes.

On the right, in the hedgerow that flanks the path, Song Sparrows and Bewick's Wrens are in good form. They sing loudly from perches chosen for reasons known only to themselves. A Fox Sparrow, by contrast, seems content just to call at us with his husky *chup*. Overhead, a pair of Killdeer flies in wide circles, one, perhaps the male, calling constantly. In the distance, two male Ring-necked Pheasants have squared off in a verbal sparring match.

Skirting the end of the hedgerow, we flush a large group of Northern Shovellers from the canary grass; they scatter, regroup and settle a little farther out in the flooded field. Diminutive Green-winged Teal watch us furtively as they paddle behind tufts of grass. Mike indulges me as I do a quick count of the waterfowl; this is an important winter habitat for birds, and documentation can only be done when the birds are there. Glassing the flock, I estimate the size of a group of 10 birds, then extrapolate to 100, and then count by groups of 100: 1,100 ducks, more or less — ducks that are dependent on these seasonally flooded fields.

The ducks are not alone in the marshy scene; dotted evenly throughout are a dozen or so male Red-winged Blackbirds, hollering at each other from the enclaves they have claimed. The sun, though warm as spring, is still winter-low in the sky, and when the Red-wings display their trademark epaulettes, the fiery flash is nothing short of startling.

In the large field that opens to our right, a flight of about 15 Western Meadowlarks sweeps in to take cover in the grass. A lone meadowlark sings now, from another field. They formerly nested on Vancouver Island, but now we hear the song only briefly each spring. The meadowlarks stay down, and we soon see the reason: a male Northern Harrier flying low, quartering the field. He is dressed in pale grey and white with jet-black wingtips, and it occurs to me that all of the fashion mavens of New York and Paris have yet to match the elegance of this outfit.

From the woods to the south, a Red-tailed Hawk powers directly toward the harrier, and we watch to see who will elect to graciously leave the area. But the two raptors do a perfunctory double-feint and then apparently agree that this is too good a day to waste in fighting; they go their separate ways. Now a Sky Lark, our Victoria specialty, with the sky to itself, spirals upward in ethereal song. Its notes roll across the spring air, minute after amazing minute, until finally it drops again to the field.

We reluctantly make our way back to the car, feeling a little intoxicated by a beautiful day and a profusion of birds, all deeply involved in the business of being birds. We've seen nothing unexpected or rare, but it is a rare morning indeed; I don't think even a Painted Bunting could have made it more memorable.

Naturescaping

At this time of year, with those delicious seed catalogues arriving in the mail and the nurseries moving winter heathers aside to make room for rafts of annuals, our long-lost gardening genes begin groping for the light. The urge is happening even in me, so I know it's fairly bursting at the seams in other people, and now is the time to plant a seed of a different kind in your fertile minds.

Our gardens are key players in the growing movement to provide more "green spaces" and "green belts" and "greenways." The idea is to recognize that we must preserve some patches of green, partly for their visual appeal and partly as a break in the urban sprawl. There is a natural belief too, that as long as there is some greenery around, wildlife will automatically follow. We soon find, though, that manicured parks and verdant golf courses just don't provide the diversity of wildlife we expected, and certainly not what we find in a more natural area. Perhaps there's a clue here: Perhaps there is too much "green" and too little "wild" in these oases we have so carefully set aside.

There are encouraging signs of change, in our resistance to the threatened loss of some of our wild spaces, and our recognition that natural areas have an important place within the manicured confines of our public parks and gardens.

All this is very nice, you say, but what's this about a different kind of seed? Well, you too can play a part in providing a little wild space for wildlife. Best of all, you can do it in your own backyard.

Creating wildlife habitat where you live may sound like a lot of work, but often it's more a matter of what not to do than what to do. If you force yourself, for example, to allow those brambles to grow up and over the compost box instead of cutting them back, you may find finches feeding on the fruit this fall. If that back corner under the Grand Fir won't grow anything anyway, why not plant some native bunchberry and let it slowly take over there? It provides good cover, and the berries are eaten by birds. You don't ever have to mow it, either. A little bit of native shrubbery and honeysuckle along the fence will attract many more birds than do azaleas and bark mulch, and they also require a lot less maintenance.

Local nurseries are becoming good sources of information about using native plants, and about gardening for wildlife. Some make it their specialty; they can help you to naturalize and indulge your penchant for unusual plants at the same time.

There are several excellent references which will help you first to discover some of the amazing native plants we overlook, and then how to incorporate them into a garden that is more beautiful, more drought-tolerant and more disease-resistant than conventional gardens. A little more to the point, perhaps, are books that deal specifically with gardening for wildlife. Both native and non-native plant species can be combined to attract animals and birds to our yards, and in some cases provide the right environment for them to take up residence. One of the best sources is the marvellous Naturescape program of the B.C. government, which produces helpful books for most regions of the province. *The New Gardening for Wildlife*, by Bill Merilees of Nanaimo, is a wonder in the amount of useful information it contains, covering wildlife gardens in all four seasons. Both of these references are equally useful in adjacent areas of Washington.

Gardening for wildlife provides food, shelter and nesting opportunities for many species of birds. People are often surprised to find House Finches raising young in hanging flower baskets. Rufous Hummingbirds nest happily in a variety of tree species. Shrubs and hedgerows offer good cover for Spotted Towhees and Chipping Sparrows.

Birds fall into several categories when it comes to nesting. Some lay their eggs in a simple scrape on the ground, like Killdeer. Some nest in burrows, like Northern Rough-winged Swallows. Almost a quarter of all birds nest in cavities in trees. Some are excavated by the birds themselves, like the woodpeckers. Other species borrow the cavities when the woodpeckers are finished, such as Violet-green Swallows. Often, natural cavities are used, too. Most songbirds build their own nests, though, using a variety of materials. American Robins use grasses and sticks bound together with mud. Barn and Cliff Swallows fashion mud cups or gourds, respectively. An amazing variety of natural materials is also used: grasses, twigs, lichens, mosses, spider webs, feathers, animal hair and more.

You can see that to attract birds to your yard, it helps to know what they are looking for. When you provide good habitat, many species will nest in your garden, but sometimes you have to go a step further if you really want results. This calls for a little construction.

Nestboxes (you can call them "birdhouses," but the birds seldom do more than nest in them) are relatively easy to build, don't have to be fancy and are readily adopted by many cavity-nesting species.

The size of the box isn't critical, but it should fit the bird you're trying to attract. Nestboxes for Violet-green Swallows should have a floor area about 125 mm square. The box can be a little smaller for chickadees and wrens, and a little bigger for Downy Woodpeckers. The hole size varies with the size of the bird, too, but it is more critical in deterring non-native species which compete for the nest sites: House Sparrows and European Starlings.

Wrens and chickadees can use a 28-mm hole and that keeps most House Sparrows out. Thirty-two millimetres is good for Violet-green or Tree Swallows, but the aggressive House Sparrows will usually out-compete them, sometimes even destroying swallow eggs or young in the process. Victoria naturalist Darren Copley has found that sparrows are deterred from using swallow boxes that have oval holes, about 23 mm high and 38 wide. It's a squeeze for the swallows, but they seem to manage fine, perhaps because of their very small feet and legs.

Screech-Owls and Northern Flickers will use nestboxes and if, like Darren, you're lucky enough to have Wood Ducks around, you can even build boxes for them, too. There should be a book in your library with size guidelines for these and other less common cavity-nesting species. The Naturescape series and Merilees' book also have excellent information on nestboxes.

You can decorate the nestboxes if you like, but don't use toxic finishes and don't use anything on the inside. Don't use perches or other adornments that give predators something to hold on to — snakes are known to remove eggs from nestboxes, and squirrels and raccoons can reach in and take eggs or young too.

What about other birds? American Robins can sometimes be encouraged to build their nests on shelves erected for that purpose. Barn Swallows have a fondness for electric-light-fixture boxes, but with an artificial shelf you may succeed in getting them to nest in a place which is less directly over your back door or Mercedes-Benz.

You can also help nesting birds by providing construction materials for them. Many birds will use bits of straw or grass left on the ground. Orioles like pieces of cotton string for their carefully woven pouch nests. Many species line their nests with feathers — swallows will sometimes even take them from an upheld hand.

So, when you are landscaping or gardening, don't be too quick to tidy your yard into avian sterility. Leave a corner wild; use native plant groupings; cut down on black plastic and bark chips. Leave a dirty puddle or two for the mud-nesters. Avoid chemicals.

Birds are looking for places to raise their families. They're looking for habitat — healthy habitat. By providing them with opportunities, we help the birds, and in return, they help us with pest control and with the simple joy of their company.

Maybe This Will Be the Year

Maybe this will be the year. Perhaps it will finally be my turn. After all, I've waited long enough. Sometimes I think everyone in the city has done it, but not me. So maybe, this year, I will be lucky enough to find a Rufous Hummingbird nest.

Well, okay — perhaps "everyone in the city" was pushing it a bit. But still, I can't count the number of times people have told me about their discoveries. People find the little walnut-sized nurseries outside their kitchen windows, overhanging their patios and at eye level beside their decks.

Friends have reported hummer nests in the same tree on their property two years running. One family watched a hummingbird family in a fuchsia basket. And local birder Bryan Gates found the first Anna's Hummingbird nest in Victoria from the rocking chair in his den. But they continue to elude me.

I have seen lots of hummers, of course. Like many people, I feed them near the house during their brief summer stay here. And they have come, regularly and trustingly, to my offerings. The females even bring their young to the trough, to carry the tradition on to the next generation. This year, I am ready for them again. I have set out my feeders for the earliest males, who arrive about the middle of March; the females will appear a little later. In the beginning, at least, the males tend to dominate the feeders.

I spent quite a lot of time one year, carefully tracking the movements of a male which had laid claim to one of my feeders. I followed him until I lost him, and then waited until I saw him coming along his route again, and followed him some more. He led me to a flowering currant bush along the driveway, to a dead branch on a pin cherry tree where he had a good view and to another red plastic hummingbird feeder. But no nest. He didn't even introduce me to any of his lady friends.

I was a little chagrined to discover that I could have done better things with my time. Male Rufous Hummers do not assist with the construction of the nest. In fact, they do not even know where the nest is. The female alone chooses the site and builds the nest. She then seeks out a male on his territory, where courtship and copulation take place. From here on, there is no further contact. The female builds the nest, lays her two tiny eggs and raises the young entirely on her own.

The male continues to defend his personal feeding territory, whether it is a salmonberry bush or a feeder, and will attempt to repel anyone else who attempts to use the resource. This aggressiveness is aided by superior agility in the Rufous, and particularly the male of the species. He has a higher wing loading, which means less wing area per unit of weight, and the result is greater manoeuvrability.

For this reason, it is a good idea to put out more than one feeder; an ornery male has a hard time defending both. With luck, females in the area will be able to use the other feeders.

There are many things you can do to attract hummers to your yard. The easiest is to put up a feeder. There are dozens of commercially available models and they all work on the same basic principle. Some drip less than others, some have perches, some are easier to clean and you can choose between plastic and glass. Your local nature store is a good place to compare them and seek advice about the relative merits of each.

The nectar solution should be one part white sugar and four parts water. Boil the mixture for a few minutes to prolong its shelf life. Do not use honey or brown sugar which, although healthy for humans, may ferment or cause fungal growth in hummingbirds' throats. Do not add anything else. Food colouring is not recommended as there is still some debate about its potential to cause harm to these tiny birds. Most feeders have a splash of red colour on them anyway. Commercial nectars will work, but are no better than white sugar, which very closely approximates the sugars in plant nectar.

While feeders will get you started, you can do much more for the hummers by creating a habitat to their liking. Plant flowers that they prefer, like currants, honeysuckles and hardy fuchsias. Salvia and bee balm are favourites too, and there are dozens more. Hummingbirds are beautiful to watch at a feeder, but as they nimbly adjust to garden flowers, hanging in the air at every angle, they are simply magical.

Perhaps the most rewarding thing to do for your hummers is to provide a habitat suitable for nesting. Early in the year, they seem to prefer to nest fairly low on thick branches of coniferous trees. Many birds raise more than one brood, and later nests are sometimes a little higher, in deciduous trees.

Here is my problem: Having done all this, I am unable to determine whether I've been completely successful. The hummingbirds are there, all right, but my success would be complete if I could find a nest, just so I'd know. Maybe this will be the year.

Enjoy a Fine Selection of . . .

There are a lot of great places to go birding, but the other day I was reminded of just how great one of my favourite birding spots can be.

There are tall forested hills on all sides, sliding steeply into the steel-blue Pacific. In places, ancient rock laid bare by the glaciers of 15,000 years ago suffers yet more abuse with the pounding of salty surf. Underfoot, a thin layer of cold steel gives me an elevated perch with a panoramic view. Best of all, this birding spot moves along at a respectable speed, providing a continuously changing scene.

I love birding from ferries. Big ones, small ones, it doesn't matter; they all have one thing in common: They go from one bit of land to another. And there is always something to see en route. Many of the small ferries operate in protected waters, bays and passages sheltered enough to attract many species of seabirds. From the car decks of these vessels, a birder can have excellent looks at scoters, and those Long-tailed Ducks that are so often just out of visual reach. I recall one Saanich Inlet crossing that was speckled with over 800 Common Murres.

In northern British Columbia the run from Prince Rupert to Skidegate, in Haida Gwai'i, is more of an adventure. Hecate Strait is a shallow body of water and rough seas often delay sailings for many hours. But once out in the strait, a birder can be kept moving, from port to starboard, sorting through hundreds of Sooty Shearwaters to find other pelagic species. Northern Fulmars are there, and Fork-tailed Storm-Petrels and other species, depending on the season.

The run between Swartz Bay on Vancouver Island and Tsawwassen, on the mainland, is probably the most heavily travelled ferry route in B.C. It offers excellent birding, passing through some of the prettiest scenery on the south coast, with good birding through most of the year. The western (actually southern) end of the trip winds through small islets north of the Saanich Peninsula, where there's a good chance to see Pelagic Cormorants and diving ducks like scoters and goldeneye. A little farther out, birders can look for several species of alcids. These are efficient diving birds with pointed bills, which are so well adapted to life in the sea that they actually fly under water.

From mid-summer through the following spring, there are lots of Common Murres. They're fun to watch in March because some individuals are already wearing their breeding black, while others are still trying to get another month out of their winter coats. Pigeon Guillemots are around all year, and nest on many of the offshore islets. In recent years, they have also moved into hollow girders at the ferry docks. They can be hard to spot on the water, but as soon as they dive, their bright red feet flash like a beacon to identify them. Smaller Marbled Murrelets and occasionally Ancient Murrelets can be found along the way. Rhinoceros Auklets, apparently immune to the subtleties of international diplomacy, happily haul fish from Canadian waters back to their nest burrows on the American side. Once in a long while, a Tufted Puffin will shoot past, perhaps one of the few which occasionally nest on Mandarte Island.

Orca move through sometimes, and are usually announced by obliging officers on the bridge. You may also see Dall's Porpoises riding the wake of the ferry.

The open water at the northern end of the route offers smaller numbers of birds, but here you have a chance of seeing a Parasitic Jaeger harassing terns and smaller gulls. But the best part of the trip is when the master threads the needle through Active Pass. There is seldom a day when nothing is happening here.

Spring is a good time, with Steller's Sea Lions often hauled out on the rocks at the western entrance. Large flocks of Brandt's Cormorants, with white breeding plumes on their necks, are joined by hundreds of Pacific Loons and Bonaparte's Gulls, marshalling for the last leg of their northward migration. Careful birders scanning these flocks sometimes report much less common species, like Sabine's or Little Gulls.

One of the highlights of Active Pass is the near certainty that you will see a Bald Eagle. Except for a short period in the early fall, there is almost always at least one soaring over the rocky bluffs. You can scan the trees along the shore for white heads, dotted like so many Christmas ornaments amid the green. Active nests can sometimes be spotted along the route, too.

And once in a while, you just get lucky. On one otherwise ordinary trip, as the ferry approached the middle of the pass, I moved to the rail to glass a feeding flock which was working a "ball-up" of small fish. A feeding flock like this can develop with amazing speed and, as I watched, hundreds of birds of many species descended on the scene. There was so much food available that the cormorants and murres were completely unconcerned by the presence of at least 25 eagles, wheeling and diving with the rest.

On wings that would do an Airbus proud, the eagles cruised low over the boil, lowered their landing gear, snatched fistfuls of silver booty and pulled up. Several missed on their first grab and tried again, on braking wings, until their airspeed fell below the stall point and they were forced to power away again. One immature bird actually settled on the water, the largest "hook-billed duck" I have ever seen.

The ferry churned past, the ball-up subsided and the birds thinned out as quickly as they had collected. I could see half a dozen eagles still in the air, but the rest were out of sight. In all, it probably only lasted about two minutes.

Every passage is different, of course, and some are much less productive than others. But like most birding, just the prospect of a rare bird, or migrating flocks, or unusual events is a big part of the appeal. So I don't mind riding the ferries at all; it's one more great place to go birding.

The Emperor and the Kelp Bed

Nineteenth-century French politics would seem to be worlds away from a kelp bed on the coast of British Columbia, but even here, there is a connection.

In 1803, Napoleon Bonaparte who, three years earlier, had installed himself as dictator of France, completed one of the largest real estate deals in history: He sold the Louisiana Territory to the fledgling United States of America. In that same year, a son was born to Napoleon's younger brother, Lucien. The son's name was Charles Lucien Jules Laurent Bonaparte.

The eldest of 11 children, Charles Bonaparte was born in Paris, but returned to the family's homeland of Italy to complete his education. Although he inherited the title of Prince of Canino and Musignano, he was educated not in politics or the military, but in science. His decision may have been influenced by his uncle's defeat at Waterloo and subsequent exile to St. Helena.

At the age of 19, Charles married his cousin Zénaïde, the daughter of the Emperor Napoleon's older brother, Joseph. The two left almost immediately for America, along with Zénaïde's sister and father, and they all settled in Philadelphia.

Charles Bonaparte was not a dilettante naturalist. In Europe, he had discovered a species new to science, the Moustached Warbler. On the voyage to America, he collected specimens of a storm-petrel which he later named after the eminent ornithologist, Alexander Wilson. His accomplishments secured him a position at the Philadelphia Academy of Natural Sciences, where he set about completing a catalogue of North American birds that had been begun by Wilson. In the end, he spent only eight years in the United States, but he is credited with establishing a systematic ornithology for North America.

As was the fashion in those days, he sometimes honoured his wife when he was cataloguing the North American avifauna. He named a genus of doves after her, which, Latinized, became Zenaida. One species of the genus, found only in the Yucatan Peninsula, also bears the common name Zenaida Dove.

It is not permitted (nor is it politic) to name a species after oneself, but Bonaparte is honoured for his work in having a species of gull named after him. The common name is Bonaparte's Gull; the Latin or scientific name is *Larus philadelphia. Larus* is the genus name for almost all gulls, and *philadelphia* refers to the city near which the type, or first, specimen was taken. The honour was bestowed by George Ord, a Philadelphia scientist and another close follower of Alexander Wilson.

Bonaparte must have been pleased with his namesake. In a family that includes large, aggressive scavengers, his name has been given to one of the prettiest and most delicate of the gulls.

Bonaparte's Gull is one of many gull species worldwide that are quite small and reach full adult plumage in only two years, unlike the larger gulls. In breeding plumage, this bird stands out, its gull-grey upperparts offset by a striking black hood. The bill is small and black, and the legs and feet are red. Add to this long, pointed wings, which are tipped black, and you have a very elegant little gull.

In flight, Bonies (perhaps Bonaparte would not be too pleased with this nickname) show a distinctive white triangle in the fore portion of the outer wing, and their size and delicate flight are reminiscent of terns. Their calls are much softer and less strident than the larger gulls, and include a kind of guttural gabbling.

While most gulls nest on the ground, Bonaparte's are a little different. They nest in a tree, up to five metres off the ground, and generally close to a lake or pond in a coniferous forest. Some years ago, I was struck by how odd it seemed to see Bonaparte's Gulls on territory at a small lake surrounded by Subalpine Fir in Tweedsmuir Provincial Park in west-central British Columbia.

The breeding range covers much of the northern two-thirds of the province, but in the south they pass through twice a year, during spring and fall migrations. Large numbers move through in April, with concentrations of up to 10,000 at productive feeding sites like Active Pass and Porlier Pass, in the Gulf Islands. They follow the tides back and forth in orderly groups, their slim wings carrying them with graceful aplomb just above the wave tops.

In the fall, they return south along the coast; the juveniles lack the black heads of the adults and have mottled upperparts. When the termites take wing, the Bonies are manoeuvrable enough to catch them, flycatcher-style, in a most un-gull-like fashion. They like to loaf on kelp beds along the shores, sometimes joined by other black-headed gulls — Franklin's, from the prairies, and occasionally Black-headed and Little Gulls, wanderers from Asia.

And there lies the connection between Waterloo and a B.C. kelp bed. You're not likely to read about it in a book about French history, but it's a pretty good story, just the same.

Bonaparte's Gull

Townsend's Solitaire

April

Ogden Point

The long concrete breakwater at Ogden Point in Victoria is an unlikely-looking friend to birders. Its windswept, unvegetated expanse supports few birds, although a few shorebirds can usually be found on the wave-splashed granite blocks below (birders call these birds "rockpipers"). It really comes into its own, however, as an observation platform. With its long reach into open water, it offers good views of several alcid species and occasionally more pelagic species, such as Northern Fulmars. On one occasion, a Kittlitz's Murrelet from Alaska was discovered here. Another vagrant from another ocean appeared here in April; a Scissor-winged Storm-Petrel was seen briefly off the point and then, as often happens, it vanished.

This is one of those far-flying seabirds that seem hopelessly lost, and thousands of miles from home. It is probably only because of the extraordinary endurance of pelagic species that it survived. Normally found in the warm waters south of Japan, Scissor-wings disperse for the non-breeding season in a broad, clockwise sweep through the South China Sea, and west to the Indian Ocean. This bird apparently dispersed in the wrong direction and, following the nutrient-rich Japan Current, found its way to North America.

Until recently, the Scissor-winged Storm-Petrel was considered to be a subspecies of the very similar Swinhoe's Storm-Petrel, although Robert Swinhoe himself recorded the two birds as separate species in the 1860s, based on the structural difference he observed in the wings of the two. While Swinhoe's has a conventional high-aspect wing typical of pelagic species, the Scissor-wing is quite unique.

In this species, the *alula*, or false wing (the "thumb") has dramatically longer feathers on it, and this makes the wing look like a pair of scissors; hence the common name. It became apparent that the two species were interbreeding freely, and thus were considered conspecific, and for the next 120 years or so that was the way things sat.

Research done in the late 1980s by a British geneticist, however, has determined that the offspring of hybrid pairs of these two birds invariably have the conventional Swinhoe's wing; only the offspring of paired Scissor-winged birds will produce offspring with the extended alula. This means the chromosome which determines the length of the alula must be carried by both parents. As a result of this research, the Scissor-winged Storm-Petrel has again been given full specific status. According to the system of taxonomic priority, it keeps the scientific name given to it originally by Swinhoe himself.

The research also confirmed Swinhoe's observation that during moult, a Scissor-wing will occasionally lose the long feathery alula on one wing before the other. The missing feather soon grows in again, but while the two wings are in a state of airfoil imbalance, the bird is compelled to fly in large circles. For some, their altered flight carries them away from their normal route, and into the path of the Japan Current. Then, swept along over the Japan Current, they end up on our shores. Today's report of a Scissor-winged Storm-Petrel is certainly unusual, but it does follow a distinct pattern; over the years all North American sightings have been made on the morning of the first day of April.

Concan, Texas

We enjoyed many rewarding birding spots on a Texas trip. Along the Rio Grande (which is not so grande), birds more normally found south of the border, like Green Kingfishers and the surreal Green Jays, greeted us on the American side of the line. White-tailed and Harris's Hawks, and a surprise American Swallow-tailed Kite soared overhead on one of the sunnier days. We watched as a kettle of 1,300 White Pelicans peeled off in smaller groups, following their instincts northward.

We saw birds in ones and twos, like the Clapper Rail at Galveston, and we saw a field seething with Whimbrel. Thousands of American Avocets festooned the waves at Bolivar Flats. The most dramatic spectacle of all, though, involved millions of wings, but only a relative handful of feathers.

Near the little town of Concan, west of San Antonio, there is a famous cave, which has been owned by the family of Irving Marbach for many generations. It is renowned as the daytime roost of several million bats, of two species, and is called the Frio Bat Cave.

We were taken to the mouth of the cave by Mr. Marbach early one evening, while the sun was still well up. A Canyon Wren sent its cascading song ricocheting off the rocky hills. Overhead, several hundred Cave Swallows gathered, chattering constantly, preparing to enter the cave for the night.

The sun fell lower and the smell of the bat excrement began to waft out of the opening, pushed by the wings of the bats deep inside as they began to move. A few minutes later they began to fly out, first in a trickle, and then a steady stream, rising and heading off to the east. The numbers swelled and legions of the small mammals rose skyward, looking like plumes of smoke roiling toward the horizon. They poured out, non-stop, and in my awe I lost track of time. I cannot tell you precisely how long the bats came from the cave; in retrospect I would guess between 20 and 30 minutes. The sheer number of individuals is staggering.

Humans were not the only creatures that were aware of the phenomenon. Overhead, Red-tailed Hawks took turns diving into the swarms of bats and they were sometimes successful in catching one.

These were Mexican Free-tailed Bats, and the number using this cave is estimated to be several million. They leave the cave early and fly quite high, travelling to feed particularly on moths that frequent corn fields.

The flight of the Free-tails finally slowed to a trickle. As dusk began to fall, a second species, the Cave Myotis, made its appearance. These bats normally leave through another opening and fly closer to the ground. We stood near the maw of the cave, human boulders in a torrent of bats, the soft sounds of their wings surrounding us as they streamed past. They followed the lay of the land, flying downslope and spreading out more quickly than the Free-tails over the dry, shrubby countryside.

Over the years, the Marbach family has made a bit of a living from mining. The ore? Bat excrement, or guano. Before the advent of chemical fertilizers, guano was a valuable commodity. Today, Irving Marbach still has a crew which removes the bats' excrement, which is becoming prized for organic crop production.

More recently, there has been the revenue from naturalists, wanting to visit the Frio Bat Cave. Marbach, I think, welcomes the supplementary income, but he is also genuinely interested in his winged tenants. He cooperates happily with researchers, and is very knowledgeable about the bats himself.

Tonight, in the hill country of Texas, the bats will fly again, as they have for who knows how many generations. I will not be there to watch them, but the spectacle will not soon fade from my memory.

The Gentleman in the Grey Flannel Suit

Many years ago, on a warm day in mid-April, I met a distinguished-looking gentleman in a grey flannel suit. At least, I assumed he was a gentleman. We were not introduced and we did not speak, but we watched each other off and on for the better part of a day. He was in the top of an old Garry Oak, while I was confined to the ground beneath him.

That was my first encounter with a Townsend's Solitaire. As a relatively new birder, I was a little unsure of what I was seeing. Solitaires are understated birds, almost nondescript, and I was struggling a little with the identification, looking for the definitive field mark. But this cooperative fellow gave me plenty of time to go back to my old Peterson and sort out the subtle marks of this species I had not known before.

Since then, I have always thought of solitaires in April. And April is a good month for them, because they are on the move northward, and are reported regularly in small numbers as they pass through the Georgia Basin lowland.

Solitaires are members of the thrush family, most closely related to the bluebirds. They are primarily insectivorous and, like the bluebirds, have smaller, slimmer bills than the American Robin and some of the other thrushes. Solitaires tend to perch in the open, and their comportment is quiet, dignified and, well, gentlemanly.

Townsend's Solitaire is the only species of solitaire in North America. There are two more in Mexico and others farther south. In Mexico, I have often seen Brown-backed Solitaires for sale in the market; they are beautiful singers, as are Slate-colored Solitaires, but they do not enjoy the protection that songbirds do in North America.

On this coast, where we see the Townsend's Solitaire primarily in migration, we seldom hear its song. In my estimation, it rivals the much-touted Hermit Thrush for the quality of its singing. Farther north, on a trip out to Bella Coola, I watched one at close range, and listened breathlessly as it spilled its rolling cascade of notes out over the timbered valley.

This gifted songster, however, is easily overlooked if he is not singing. A little smaller than a robin, he is soft grey above and a little paler below. His outermost tail feathers are tipped and edged with white that flashes in flight. The small bill is black, and the dark eye is highlighted with a thin white eyering. There is a rather incongruous buff patch in the wing that is easily missed; in flight it shows up much better. And if "he" is not singing, he just might be a she, because males and females are similar in appearance.

Solitaires are rather aptly named birds; they seem often to be alone. Townsend's Solitaire also has a particularly lovely scientific name: *Myadestes townsendi*. The first half is from two Greek words meaning "fly" and "eater"; somehow the Greek is more charming. And even Townsend's name has a pleasant ring in its Latinized form.

For a bird that breeds across much of northern British Columbia, little is known about the solitaire's nesting behaviour. Typically, they nest on or near the ground, in a somewhat untidy structure. Most B.C. nests are reported from cutbanks in forested areas. Three to five eggs are laid, white with a pattern of brown spotting. The incubation period is not known, nor is the age of fledging of the young.

Look for solitaires as they move north in the spring. They are usually found in high open situations but they may turn up anywhere. One April in Port Alberni I found one on the lawn at a marina, two metres above sea level. They also occasionally winter in southern B.C., where they take advantage of winter berry crops.

I have a particular fondness for grey birds and this is one of my favourites. While the spring migration brings many beautiful birds, April is just not April without a Townsend's Solitaire.

John Kirk Townsend

In the early years of the 19th century, the natural history of western North America was no longer a complete mystery. Still, the thin lines drawn across the maps by the explorations of Lewis and Clark, and Mackenzie, and Franklin, had but scratched the surface of that vast frontier. For contemporary naturalists, the untrammelled west continued to offer great opportunities for discovery.

John Kirk Townsend was a naturalist who saw the possibilities in the west. Born in Philadelphia in 1809, Townsend followed an early passion for birds. In 1823, he collected a bird which to this day is the only known specimen and remains a mystery; perhaps it was a hybrid, but also perhaps the last member of a species now extinct.

Townsend was eager for adventure and discovery and, through the influence of his friend, the botanist Thomas Nuttall, he joined the Nathaniel Wyeth expedition to the west coast. On April 28th, 1834, the expedition set off across the prairies of northeast Kansas and eastern Nebraska. Wyeth led a large group of traders, missionaries and others, with some 250 horses.

Lewis and Clark some 30 years earlier had followed the Missouri River system west to the Oregon Territory. Nuttall and Townsend collected intensively as the Wyeth party moved instead up the valley of the Platte River. Chestnut-collared Longspurs and Lark Buntings had never before been recorded by scientists. The same was true of the Mountain Plover and Common Poorwill, whose scientific name honours Nuttall: *Phalaenoptilus nuttallii.*

The collecting continued at a hectic pace, with the thrill of new discoveries every day. In the mountains of Nevada, Wyeth established a trading post, and here the expedition broke into several groups. The two naturalists continued westward, but endured considerable hardship in the high, rugged country. Food was running short, and they moved as quickly as they could to reach the coast via the Columbia River.

They stayed through the fall and, after a voyage to Hawaii, returned to the Pacific Northwest to continue collecting. Townsend secured specimens of two new warblers, which were later named the Hermit Warbler and the Townsend's Warbler. He also shot what seems to be the only Townsend's Solitaire he ever saw.

Nuttall prepared to leave the area for home, via Hawaii, and Townsend elected to ship his bales of specimens home in Nuttall's care, which left him free to continue his explorations. When Nuttall arrived back in Philadelphia, word of Townsend's collection had preceded him, in particular to John James Audubon who was desperate to add the new western species to his work on North American birds.

After much wrangling, Audubon persuaded the reluctant caretakers of the collection to part with it, paying for the specimens with cash borrowed from a wealthy patron. Audubon professed to have no interest in the matters of priority, and agreed that Nuttall should write accounts under Townsend's name. Nuttall, perhaps preoccupied with botany, did so, but provided only brief descriptions of 12 species.

In correspondence with friends, Audubon gloated over the wealth of specimens he had acquired, and at a very low price. He went on to publish information on many species, thus taking credit for their "discovery." Fully one-seventh of the birds in his landmark work, *Birds of America*, came from the Townsend collection.

Townsend returned from his travels impoverished and in debt. He was, however, given a position at the Academy of Natural Sciences and wrote an account of his travels in the west, which remains a classic of early collecting accounts.

The relationship between Townsend and Audubon is confusing. Townsend was uncooperative in some of Audubon's requests for material, but used Audubon to sell parts of his collections to buyers in Europe. It seems that there was no love between them, but perhaps in the small community of ornithologists in North America at that time, a certain amount of contact was unavoidable.

Townsend was regarded by his colleagues as a brilliant scientist, careful and thorough. He publicly noted several mistakes in Audubon's work, including misnamed species and errors of fact. Audubon misidentified a MacGillivray's Warbler as a Mourning Warbler in one work; when he later corrected his error, he used a name of his own choosing to identify the "new" species, ignoring the name assigned it by Townsend when he had discovered the species. Townsend had named the species *Sylvia tolmiei* in honour of William Fraser Tolmie, a Hudson's Bay Company factor who later became a prominent figure in the young city of Victoria. The species is now listed in the genus Oporornis, but Townsend's recognition of Tolmie in the species name has survived Audubon's indiscretions. His skill in preparing study skins was well known. In one instance he completed in five minutes a job which my own clumsy fingers have required two hours to finish.

John Kirk Townsend died in 1851, at the age of 41, his death possibly a result of poisoning from the arsenic dust used in the preparation of thousands of bird and mammal skins. Although he accomplished much in his travels, he never reached his full potential as an ornithologist; he was pre-empted by Audubon of the opportunity to write up his discoveries in the northwest as a single account, which would have given him the recognition he deserved. He also lacked the financial backing that Audubon seemed able to attract. Today, it is Audubon who is regarded as the father of North American ornithology.

Townsend's name lives on, though, in the scientific and common names of a solitaire and a warbler, both species unique to the western fauna he helped to discover.

Choose Your Partner

As the days lengthen in spring, birds begin to prepare for the breeding season ahead. The longer days stimulate the production of hormones, and these hormones in turn initiate remarkable changes in the birds, both mental and physical, that allow them to put their reproductive urgings into action.

In their remarkable adaptation as lightweight flying creatures, birds have evolved reproductive systems that shrink to almost nothing in the non-breeding season. With the release of various hormones which control the development of the reproductive organs, the testes in the males and the ovaries in the females become dramatically larger. Hormones also begin to influence plumage patterns in some species, and the urge to find a suitable mate becomes a high priority.

In order for successful breeding to occur, one critical condition has to be met: two birds, a male and a female, have to form a pair bond. And in birds, this process of pair bonding is almost as varied in its forms as are the birds themselves.

Just about everyone is familiar with the longevity of the pair bond in Canada Geese. There are reports of birds that remained paired for 42 years. However, this sort of relationship is in fact not the norm; it is an extreme example of what is known as a "monogamous relationship." Monogamy in birds, strictly speaking, means a pair bond which lasts through at least one breeding cycle, although some birds may be "monogamous" while they raise one brood together, and then find different mates for a second brood in the same season. Monogamous bonds can last for two or more years in songbirds, but this can vary even among members of a single species.

Monogamous relationships are formed in the vast majority of bird species, although they may last for one breeding season or less. DNA research has now shown that even in these apparently stable bonds, some birds engage in what is called "extra-pair copulation." This behaviour can help to ensure that a female is successfully fertilized, or that she is fertilized by more desirable males than the mate she has chosen.

Some birds form more than one pair bond, a situation known as polygamy, but here the semantics get a little complicated. If a male bird mates with two or more females at one time, and defends the territories in which those females nest, the bond is said to be polygynous. If it is the female who bonds with several males, then the situation is known as polyandry. This is found most often in species such as the phalaropes, in which the males incubate the eggs and care for the young.

There is evidence, however, that in some polygamous bonds the bird that stays at the nest may have more than one mate. An example would be a male Red-necked Phalarope who mated with more than one female, each of whom would be expected to mate with several males.

This lack of fidelity is called "promiscuity." It occurs normally in gallinaceous or game birds such as Sage Grouse, in which both sexes come together only to mate, and each may mate with many birds. Sometimes a pair bond is considered to be promiscuous if the bond is formed only for the purpose of copulation and then ends, even though either the male or the female may copulate with only one bird. This occurs with Rufous Hummingbirds, as the females build the nest and raise the young entirely on their own, having bonded with the male for perhaps no more than an hour or two.

It is difficult to understand the reasons behind all this confusion, because too often we try to make sense of the behaviour of birds in human terms. For example, a male House Wren may arrive on Vancouver Island about the end of April. He immediately begins marking his territory and starts building several prospective nests for his intended. She, having accepted his invitations, mates with him. They choose one nest, finish it, and she lays seven eggs. The female incubates them for about 14 days, after which the young are cared for by both parents for about two weeks. At this point, the female leaves the family, and the male is left to continue feeding the fledglings until they can fend for themselves. Meanwhile, she runs off with another male full of hormones.

In human terms this is scandalous, but it is a very efficient use of the breeding season for House Wrens. The female, the only one who can lay eggs, is freed from the responsibilities of rearing the young of her first family so that she can proceed with a second. The survival of young birds is very low; the majority do not survive their first year. So it is a very effective breeding strategy for a female to lay more than one clutch, and it makes little difference which male raises them.

The rules of this mating game may seem complicated and perhaps even purposeless but every species has evolved so that its methods are proven successful. And every year, when a male and a female respond to the appropriate signals and form a pair bond, it is their genes that will ensure the survival of their own unique species.

Rich Little

When I am working outside around the house, I always keep my senses attuned to what birds are in the yard. I sometimes think it's a good thing that somebody is not paying me for a day's work, because a full day without stopping to query an odd raptor overhead or to pick out the dry trill of a Chipping Sparrow is a rare day indeed.

I keep track of the birds I see at home. It's partly to add to the bank of information that's building about our birds, and it's also a bit of a game. How many species today? Will the hummers arrive earlier this year? Will I add a new species to my yard list this month?

If I were in fact being paid for my yard work, my hypothetical employer would be mollified to know that most of my backyard birding is done by ear. Once you have learned some of the more common bird songs, you needn't lift your head from the task at hand to record them in your mental notebook. Lately, though, I've had to be a little more careful. It's not that I'm losing my touch; the birds are playing tricks on me.

When I hear a Killdeer calling as it passes overhead, I stop now to make sure that it really is flying overhead. As often as not these days, its call is coming from a dead branch about two metres up in a willow tree — not normal behaviour for a Killdeer.

This has been a very good branch for my yard list this summer. I've heard Spotted Towhees there, and Brown-headed Cowbirds. A Glaucous-winged Gull was a bit surprising, but not compared to the pair of courting Bald Eagles. It was a Western Meadowlark, though, that really put me on my guard.

Meadowlarks were singing near here in the spring, but they gradually stopped, having moved off into the interior, most likely. But one turned up in the yard, on this same willow branch, in fact. Meadowlark shape, long yellow bill … yellow? … speckled black plumage … black? Hmmph. Starlings again.

It's amazing how easy it is to forget some of the lessons we learn. I have heard Western Meadowlarks singing in the woods. Red-tailed Hawks have screamed at me from hardhack thickets. Brown Creepers have called from pasture fence posts. And it's all European Starlings.

Starlings are very skilled vocalizers. They have a wide repertoire of sounds of their own, but some of them are also particularly adept at mimicry. One of the first lessons birders should learn is, "Don't forget starlings."

One day, as I was walking across the parking lot at Swan Lake Nature Sanctuary in Victoria, I heard a bird calling from above and ahead of me. "Greater Yellowlegs," I thought mechanically. I looked up, and it called again. But it was not moving, not in flight; it was calling from the dense foliage of a tall tree. I never did find the source of the song, but at one point the imitation became less than perfect, and I knew that I had been fooled (again) by a starling.

We on the west coast do not hear much in the way of mimicry, apart from starlings. Steller's Jays do a creditable Red-tailed Hawk, and crows sometimes give me pause. But the best North American mimics are the thrashers, the Northern Mockingbird and the Gray Catbird, none of which occurs regularly this far west in Canada.

Ornithologists don't seem able to decide just what purpose vocal mimicking fulfils. Some feel that accomplished mimics may impress prospective mates, because they have survived long enough to learn more songs. A more prevalent view is that mimicry is simply a release for surplus energy or drive. Patterns of mimicry vary among individuals and among species as well, so it is difficult to do more than speculate on the evolutionary imperative behind the behaviour.

This particular starling in our yard seems to be thoroughly enjoying himself. He flaps his wings and sings and preens, and then sings some more. But although his repertoire is pretty varied, other starlings have shown even greater versatility. They have been heard mimicking the sounds of many species of birds (including those of other mimics). They make gates squeak, and one woman in England answered her telephone, which was not ringing (but a starling was).

Perhaps the most remarkable example of starling mimicry was an individual that could imitate the call of a Common Flicker. I've heard many starlings do this, but this bird added a twist of his own; he also hammered with his bill on his nice hollow nestbox, duplicating the flicker's drumming.

I've taken rather a grudging liking to our Rich Little of the bird world. And I've even stopped complaining about those species that keep appearing and disappearing off my yard lists. Sometimes, if you can't beat 'em, you have to join 'em; so now I have a new list: Birds That Starlings Have Imitated in My Yard.

May

Red-breasted Sapsucker

Warbler with a Secret Code

A Red-winged Blackbird is a gratifying bit of feathers. It is a black bird, and it has red on its wings. The logic in its name is unassailable. The same is true of a Yellow-rumped Warbler, which has a yellow rump. True, it's not the only warbler with a yellow rump, but a yellow rump it most assuredly has. And there is the Orange-crowned Warbler. As the name would have you believe, this bird does have an orange crown. It is even possible to see it sometimes. But as a field mark, helpful in identifying the bird, it is next to useless.

Orange-crowned Warblers are among the wood warblers; small, active birds with thin, insect-eating bills. The males of many species, in their breeding plumages, are brightly marked with prominent patches of yellow. They are birds of British Columbia summers, with all but a few individuals spending their winters much farther to the south.

In a family known for bright plumages, the Orange-crown could not be much more drab. The sexes are similar, a trait that is also unusual in the warbler family. They're olive green above and a little lighter on the rump. The underparts are yellowish, with a very few faint streaks. That's it. There are no tail spots, no wingbars, no eyerings. The face is distinguished only by the faintest of lines over the eye. And the western races are the most brightly coloured of the lot, with eastern and northern birds even more nondescript.

In describing the bird's field marks, it's safe to say that it has none. That *is* its field mark. If you take the opportunity to examine Orange-crowns which have stunned themselves on windows, there, at the bases of the crown feathers, is a suggestion of orange. It's seldom seen otherwise, but sometimes a hint of it shows in the field, late in the season, when the tips of the crown feathers become a little more worn. The Orange-crown's scientific name is *Vermivora celata*. The first part means "worm devourer." The second means "hidden" or "secret" and is said to refer to the inconspicuous crown.

Its crown certainly is hidden, even secret. The name is also appropriate, though, for a bird that disappears in the emergent foliage of trees in spring. Many times I have scoured a Bigleaf Maple in search of a bird which I can hear singing, right there in front of me, to no avail.

Most warblers, despite what the name suggests, have rather thin, buzzy songs, and the Orange-crown is no exception. It sings a high-pitched trill which either rises slightly, or trickles down at the end. It's not a difficult song to recognize, but early in the year, when the Dark-eyed Juncos begin their drier trills, things can be a bit confusing. Occasionally, too, a bird with a different dialect will pass through in migration. One bird at Mount Work on the Saanich Peninsula confounded me for a time, until a search of my bird song recordings told me it was an Orange-crown, but with an Alberta accent.

Apart from the Yellow-rumped Warblers that sometimes precede them, Orange-crowned Warblers are the first of their kind to arrive in southern B.C. On the coast, they are sometimes heard as early as late March, and they arrive in the Okanagan Valley a couple of weeks later. The males begin immediately to establish territories and attract mates in preparation for nesting.

Orange-crowned Warblers nest on the ground, usually, but sometimes use a low branch. The nest is well-concealed, finely woven of grasses with a lining of plant down and feathers. Although the nests are inconspicuous, they do fall victim to predators. Garter snakes often take Orange-crowned nestlings, and crows and ravens are ever alert to adult birds feeding young in the nest. Many nests will have three warbler eggs and one larger egg, left by a Brown-headed Cowbird. This parasitic species is a relatively recent arrival in B.C., but does not seem to be having as serious an impact on the nesting success of the Orange-crowned Warbler as it is on other species. The greatest threat to the warbler's survival seems to be loss of habitat in breeding areas and of winter habitats in Central America.

Whatever the future holds, the Orange-crowned Warblers are back with us in spring. Like the pungent fragrance of the cottonwoods, their sibilant songs proclaim the new season. An amazing little bird, this, with no field marks, except an orange crown you can't see.

The Port Renfrew Big Day

There are birding events called "big days" across North America which attract all kinds of hotshot birders. They are sponsored by big-name European and Japanese optics manufacturers, and all the birders go hell-bent for herons for 24 hours in the name of conservation and birding glory.

There is, also, the Port Renfrew Big Day.

The Port Renfrew Big Day is an annual birding extravaganza which attracts only two birders — myself and a pal of long standing, Alan MacLeod — who like to pretend we are hotshots, but start late and finish early, and stop to make sandwiches with pesto and extra old cheddar.

It is a Big Day in the sense that we do try to see as many species of birds as possible in a 24-hour period, in the area around Port Renfrew, about two hours' drive west of Victoria. Why Port Renfrew? Well, there are several reasons.

First of all, there is no pressure. Nobody has ever challenged the record (which we hold). So there is nobody to out-bird except ourselves, and even if we should fall short of the record, there's no one to ridicule us, except us.

Second, Port Renfrew is a geographically very appealing area. There are sandy beaches in the bay called Port San Juan, and rocky shores in Botanical Beach Provincial Park. The San Juan River cuts through a variety of wooded habitats before it widens, with exposed sandbars, at tidewater. Even the clearcuts are home to some species of birds. Port Renfrew is also situated so that migrant birds wandering off course might see a place to land and feed and take stock of an unexpected situation.

We are always on the lookout for rare or unusual birds on our Big Day. In most years, though, the most interesting finds are species which are relatively common on southern Vancouver Island but which have not officially been recorded at Port Renfrew. It's a chance to fill in some of the dots on the distribution maps.

We have seen, for example, Northern Shovellers on our count. That's a new record for the area, at least according to our most up-to-date sources. American Wigeon and Green-winged Teal are in the same category, and we were a little more surprised one year to find Gadwall, on salt water, which really did seem out of place for a marsh-loving dabbling duck.

There are lots of regulars around, too. The most common warbler every year is Wilson's, the song of Varied Thrush is almost everywhere and American Robins are ubiquitous. At one of our campsites at Lizard Lake, we heard Townsend's Warblers, as expected, but also Black-throated Grays, in the deciduous edges. Black-headed Grosbeaks have been surprisingly common.

A Western Kingbird on our first Big Day was a long way from its home in the dry interior, but one of our most unexpected finds was a bird that found us. As we ate lunch in a gravel pull-off in a logged area, a Whimbrel circled and landed, kept company with a pair of nesting Killdeer while we ate, and then left again. Its timing was impeccable.

There is always a nice group of birds in a marshy meadow, including warblers and Savannah Sparrows. One warbler, which had a pale grey-green body and a yellow head, stumped us completely, and remains unidentified to this day. It must have been some sort of aberrant plumage. Sometimes the mist on the meadow is parted by a small herd of Roosevelt Elk, which tickles even a couple of hotshot birders.

The weather is as variable as the habitats, ranging from wet to sodden, and as naturalists we record as much detail as we can. On one count, Alan demonstrated to me his new watch, which told us not only the time and date, but also the altitude above mean sea level. Alan set the instrument at zero as we passed the breakers at Orveas Bay, and checked as we put up our tent at Fairy Lake — 10 metres. By the time dinner was over, the campsite was at 40 metres. When we slipped into our sleeping bags, the campsite had risen to 60 metres, as the altimeter responded to plummeting barometric pressure.

Alan's watch was correct on that count, and we birded in a downpour for most of the day. But usually the weather breaks enough to let us stay relatively dry and find some birds.

So, every year, in our quiet little way, we've worked at rewriting the record books. No fanfare, and no corporate endorsements (not even from a watch manufacturer), but like true competitors, every year we look forward to the next annual Port Renfrew Big Day. We want to be ready for the competition.

Kamloops, B.C., May 1994

The banquet was over; the guest speaker had been thanked. The meal and the wine had both found their marks; people were beginning to fidget in their chairs. The final duty was to rally those who would take part in the evening field trip; meet at 9:00 P.M. in Parking Lot A.

It had been a good meeting, this fourth Annual General Meeting of the British Columbia Field Ornithologists. Most who registered had taken part in the evening social the previous night and then hauled themselves out of bed in time for 5:00 A.M. field trips on Saturday. Seven solid hours of birding in the hills around Kamloops, a quick lunch, then an afternoon of business and excellent presentations by several speakers. The banquet was a fitting finish; it didn't require much concentration after a long day.

At 9:00, the keen birders gathered for the evening outing. Rick Howie, who has almost single-handedly written the book on Flammulated Owls in B.C., was scheduled to try to find some for those who were interested.

When he left the banquet hall, Rick was blown back by the wind. Not a good start for owling, but then, you never know. Well, he would give it his best shot. If the group was small enough, they could take their time and draw the owls out.

A crowd gathered around him in the parking lot for instructions. People moved to get close enough to hear; car pools were arranged; directions were memorized. Sixty car doors slammed, and the trip was under way.

Certain conditions are required for owling; all birders know this.

Sixty people on an owling trip is the equivalent of the kiss of death. Still, there were no quitters. The procession moved out, stopping at the last critical junction. Other cars slowed respectfully as they passed the group, wondering, no doubt, why anyone would organize a funeral at night. The cavalcade climbed higher into the hills. Clouds of dust from the gravel road danced in headlight beams.

One by one, the vehicles turned into a grassy pull-off behind the leader and orderly rows materialized. At an altitude of 1,100 metres, it was chilly. As people zipped up down vests and jackets, a few raindrops fell.

By the time the last car arrived, it was raining heavily. No birder would rate owling in the rain as a good prospect and yet, no cars left. Rick explained that we should listen for the low *boo-boot* of this tiniest of B.C. owls. The group spread out, and the sounds of shuffling feet and Velcro pockets gradually stopped. The rain did not. It cannonaded off Gore-Tex hoods and Tilley hats as ears strained to pick up the calls of owls. Politeness reigned. One birder called out, "Welcome to the dry interior of B.C."

Then, there it was. Distant, difficult for many to pick up, but there for sure. Rick, committed to the job at hand, began to move into the inky black of the forest after the bird. Then another called, closer and on the other side of the road.

Rick switched his attentions to this second bird, walking down the road to locate it. The group followed when called, 120 boots crunching along the muddy gravel. The bird was quite close to the road. Many people had taken off their hoods to hear better, as other owls responded now from several directions. Water dripped off sodden hair and down necks, and even Gore-Tex let some of the downpour through.

The owls, said Rick, would be hunkered close to the trunks of trees, out of the rain. Which is the more sophisticated species, somebody wondered aloud, and a ripple of laughter went through the crowd. After a short walk into the bush, people realized that finding the owl was not likely. Many commented quietly that they didn't care; they were there, and the owls were there, and that was a rare treat.

As the birders gathered again on the road, the rain increased, and mud splashed up on pant legs. "Gary Larson should have been here," someone quipped. The group moved back to the cars, in quiet conversation, as the Flammulated Owls continued to call.

A few minutes were spent at an adjacent marsh, listening to a Sora and a Marsh Wren, singing in the pitch black; hormones are amazing things. The group decided to finish the evening at that point, and cars worked their way back to Kamloops, where the roads were, strangely and ironically, dry.

It shouldn't have worked. The weather was foul; there were too many people. But those who were there will not soon forget it. Some people will think us crazy, but some, perhaps, may want to join us next time: the British Columbia Field Ornithologists.

A Weekend to Remember

Long weekends offer a lot of opportunities to travel, play, relax or otherwise sink into self-indulgence. Many are memorable, but some really seem to stand out, and that was the case for one Vancouver Island family on a getaway weekend from their home in the Cowichan Valley. In fact, they decided to stretch it to an extra-long weekend and returned from their travels five days after they had left.

Now, when you return from a few days away, you expect to find things pretty much the same as when you left. Nobody has messed up the living room, the laundry hamper is no fuller and nobody has cleaned out the fridge. Well, our friends were in for a surprise.

When they opened the door, they were greeted by what must have been a startling sight, for their house was occupied by dozens of birds — small, sooty grey birds, with long, swept-back wings. Darting from room to room, and clinging to drapes and walls, were about 80 Vaux's Swifts.

It's not difficult to imagine the confusion as the door was opened but it didn't take long before the home-owners realized that birds do more than fly around when they're in a house for several days. The scene was by all accounts quite a mess.

These people are apparently not bird lovers, but they are certainly not bird haters. Their first priority was to open all the doors and windows to get as many swifts as possible back out where they belonged. The birds cooperated fairly well, but that was too easy. As they began to systematically clean the house, the owners came across more birds roosting in crevices and behind curtains.

One by one, they gradually released the swifts. A few, too long without food or water, were dead. Others may have killed themselves trying to fly through windows. The owners, fortunately, had the foresight to contact friends who are keen birders and as a result, the Royal B.C. Museum now has a few more good specimens of Vaux's Swifts, not an easy species to collect.

The appearance of these swifts falls neatly into the pattern of the species in southern B.C. They are seen primarily in migration in the spring and again in the fall. On the west coast, regular sightings indicate the passage of thousands of birds in September, sometimes in very large flocks.

How did they get into the house? The answer requires a little detective work. While Vaux's Swifts are not yet well known, their eastern counterparts, Chimney Swifts, have enjoyed a longer association with humans and are better understood.

Both Chimney and Vaux's Swifts have historically nested and roosted in hollow trees. Their small feet, while not good for perching, are well adapted for clinging to vertical surfaces. In the east, where abandoned chimneys became more common than hollow trees, the resident swifts took to the structures quite happily and people, having now become more aware of them, called them Chimney Swifts.

In the west, Vaux's Swifts still nest primarily in hollow trees, typically large cedars or cottonwoods, but, like Chimney Swifts, they often roost in other spots. This happens frequently during migration and there are many reports of both species preparing to descend a chimney.

They first mill around over the chimney in a general mêlée and then gradually organize themselves into a whirling column. It is as though they are compelled to enter the roost as a whole, for round and round they fly, and suddenly they began to drop straight into the chimney, one after the other, until all have been sucked from the twilight.

Down in the chimney, they cling to the walls where they can, or if it is crowded, to each others' backs. Occasionally, one will lose its grip, and flutter around until it finds a new spot.

If a gust of wind enters the chimney from either end, birds may be caught off guard and fall out of control down the chimney. It is possible that this is what happened to these swifts, particularly since the weather was quite unsettled on that long weekend.

So if you happen to read this while you're enjoying a weekend away from the chores at home, enjoy yourselves; just open the door carefully when you get home!

A Life on the Rocks

The territory was a good one. There was a nice depression on the rocky islet, where storm-driven surf last winter had washed in some fine gravel. A raised outcrop offered good views to provide early warning of predators.

The male and female Black Oystercatchers looked identical. Their bodies were a clean if sombre black but the rest looked like a collection of leftover parts from other birds. Their feet and legs were fleshy pink and their long stout bills a bright vermilion. Yellow eyes were ringed with bright red circles. The pair had courted, heads bowed, racing together along the shell beach. With the location of the nest decided, they worked on its construction: a simple scrape in the gravel and that was it, just a metre or so above the high tide line.

The female laid two eggs in the nest, as most oystercatchers do. They would hatch in about 26 days in this latitude but in the meantime, their survival depended on several factors. The eggs would require warmth for their development; she and her mate would provide that in turns. The oystercatchers would also require almost a month without above-average high tides or waves caused by wind or ship wakes. Either would wash away the eggs despite the objections of the most devoted parent.

When the dark shape of a Bald Eagle descended on the little islet, the oystercatchers shrieked in alarm. The eagle was undeterred, and landed. It sidled over to a depression in the rock and lowered its head. There, it drank from a small pool in the rock. After a few minutes it was satisfied, and left. It had not seen the oystercatcher eggs nearby, so well camouflaged were they.

The parents defended the eggs against other predators like otters and crows, and also large, two-legged mammals, humans, that left their slim plastic shells lying on the beach. The humans did not ever eat the eggs, but their appearance on the territory alarmed the oystercatchers, and they called repeatedly until the intruders left. It was only then that they could return to the nest.

The birds had abandoned one clutch of three eggs on a nearby islet the previous year. Flushed too many times from their territory by the larger and utterly fearless intruders, they chose not to risk their own safety to continue incubating the eggs. They did not renest. This year they were trying a new territory.

The oystercatchers did not have to address the question of finding mates, for this pair was in its third year together. The male had lost his first mate to a Peregrine Falcon, and courted this female in her fourth year, and first breeding season. Probably neither knew that they were one of only about a hundred pairs breeding on southern Vancouver Island.

On the morning of the 27th day at the nest, a small hole appeared in the first of the speckled eggs. Inside, the tiny "egg tooth" on the bill of the young bird worked to enlarge the hole, until it succeeded in escaping the confines of the shell. Like most shorebirds, the hatchling oystercatcher was wet, dishevelled and exhausted, but its eyes were already open, and in an hour or so, it was on its feet and peeping to its parents. Its sibling joined it in the outside world soon afterward.

The baby oystercatchers remained in the nest for about a day and a half, probably with some egg yolk still in their stomachs to sustain them. They were able to walk on their own soon after hatching, though, and quit the sparse nest to follow their parents as soon as they were mobile.

With this first critical stage of their lives concluded, the young Black Oystercatchers faced new challenges. They had to eat, to grow and to develop the flight feathers that would carry them away from danger. But that would be in July; until then, they would be confined to their tiny islet home and they would have to depend on camouflage to protect them from predators. At the first sign of danger, the adults called loudly to them, and the little birds hunkered down in crevices in the rock, where their warm down feathers, streaked and mottled, blended perfectly with the algae and lichen to conceal them.

The young oystercatchers learned that that they did not, in fact, have to catch oysters (although these were remarkably easy to catch). Small caulks on their feet helped them to grip the slippery rocks of their home as they searched for food. Their parents taught them how to use their flattened bills to pry limpets and chitons and other shellfish from the rocks, and then remove the muscular foot from inside the shell. They were shown how to thrust their bills into the open shell of a mussel and sever the muscle at the hinge. But it all took practice, and the parents helped them for a month or more.

As the young grew, and then took wing, the family joined a larger group of oystercatchers, with which they would spend the coming winter. In loose company, they would patrol the waterfront, feeding as they watched for predators.

In the spring, the older birds will begin to feel the mating urge, and breeding pairs will form. The younger birds will remain with the non-breeding flock, probably for two or three more years. When they are ready to breed, they will have to establish a territory with a suitable nest site. Some of the adult birds will not have survived the winter, so their former territories might be available. Other sites will be rejected, because of the boats pulled up on the rocks or the dogs that arrive every morning, or the people enjoying nature.

With a little luck, though, the young oystercatchers will in time find mates, and raise young themselves. Their own alarm calls will ring out across the channel, warning all that this is oystercatcher territory, and to keep clear. And with a little more luck, perhaps their calls will be heeded and perhaps their nesting will be successful.

Pages from a Birder's Diary

March 19th

The branches of the willow are still bare and empty. There are birds elsewhere: House Finches chirp in the brambles and Song Sparrows are singing from the red-osier dogwood. There are even Virginia Rails calling in the wettest of the thickets of Rithet's Bog.

A soft rush of notes comes from somewhere to the side. Then there is quiet, and still no life in the willow. The song comes again: *turr, turr, turr, turtle-y, turtle-y, turtle-y.* And suddenly, the willow is busy, not with numbers of birds, but with the movements of one tiny mite of a bird.

If I didn't know what I was looking for, I would find it difficult to see any of the features of this bird. Olive above and creamy yellow below, I can see that, but then it is gone, up to another branch. A creamy wingbar, then upside down on a catkin. Wait, two wingbars, and now it is hovering at the tip of a branch. It settles briefly, delicately gleaning food from the terminal bud, and I can see its bill, a tiny, thin, black needle of a bill.

Now it is singing again, and for a brief instant it stops moving. A shiny black eye stands out on the olive face, accented by a white eyering. And as it turns to move again, there: a small scarlet spot flares briefly on its crown. The little bird scarcely stops moving. It seems as if every branch on that tree has a bird on it, but there is only one bird to be seen, this Ruby-crowned Kinglet.

The bird's behaviour is entirely in character. Ruby-crowns are almost always alone, unlike their cousins, the gregarious Golden-crowned Kinglets. And like all kinglets, they are constantly active, always on the move, seldom stopping long enough to give a birder a decent look.

The two North American kinglets are members of the genus Regulus, which means "little king." Two related species in Europe, however, are not called kinglets; they are named after their crown flashes, Firecrest and Flamecrest. A race of the Firecrest that occurs in the Canary Islands is thought by some authorities to be a separate species, and there is also a species endemic to Taiwan, called the Taiwan Firecrest.

These birds are members of the family known as the Old World Warblers. It is a little confusing on our continent, because the North American species are unrelated to our North American warblers (the Wood Warblers), which actually don't warble very well at all. The Old World Warbler family, though, includes some very fine singers.

Most of the Regulus species have thin, lisping songs, but the Ruby-crowned Kinglet got the vocal chords in the family. The song is soft but the notes are sweet and clear, tumbling over one another. And while some birds' songs carry across the medley of spring sounds, the Ruby-crown's seems more transcendent, a part of the very air, like the smell of new growth or the mildness in spring that lingers even when the sun goes behind a cloud.

Ruby-crowned Kinglets winter in small numbers in southwestern British Columbia, but they seem to be most visible in the spring, when they are moving through on their way farther north to breed. More often than not, it is the song that says there is a Ruby-crown about. I am alert to the arrival of the Rufous Hummingbird, and the first swallows, but every year I forget how powerful a portent of spring the first Ruby-crowned song is for me.

The Ruby-crowned Kinglets will move on, but no matter; I've got the message. I can turn on the outside taps again and bring out the patio chairs. The Little King has decreed it; spring is here.

May 1st

The morning dawned, calm, dry and mostly clear, and we were a little caught off guard. Plans we had reluctantly cancelled in the face of last night's weather forecast were quickly refashioned, though, and four of us met for a long overdue walk. We set off on one of the many rough tracks that wind their way through the power-line rights-of-way on southern Vancouver Island.

These power-line trails can offer good birding, at least until the mountain bikers and dirt bikers wake up. In the fresh, cool air of this morning, the trail is alive with birds. Steller's Jays and Varied Thrushes call from the forest we pass. As we come to a marshy opening, dozens of Violet-green Swallows dart back and forth, some dodging to miss us, in manoeuvres that make *Star Wars* look like a B movie.

As the trail begins to climb, we enter the artificial "edge" habitat that is typical of these power-line cuts. Edges are favoured by a number of species that can't seem to decide whether they like it in the woods or in the open, so they go for a compromise. We hear the first spring songs of several male Ruby-crowned Kinglets, a lovely, understated song and one of the great treasures of the bird world.

I am surprised to see a trio of Turkey Vultures, aloft an hour and a half before their normal 10:00 A.M. start. They are working an updraft along a bare shoulder of rock. A Red-tailed Hawk is up, too, but keeps to its own bit of rising air.

As we pass from the trees into the clearings and back again, the birds change, too. Brown Creepers sing "*see-city, see-city*," as they work the larger tree trunks in the forest, and recently arrived Orange-crowned Warblers dribble their songs from budding Garry oak saplings in the sunny openings.

Purple Finches, more common in the woods than the House Finches we see at the feeder, are giving their hurry-up warbles from the tops of the Douglas-firs. They're joined by another bounding gang of finches that takes over the treetop level. They are Red Crossbills, coming out of a rather low year, but calling as though things have never been better.

Arriving at a high, open knob, we sit down to rest and revel in the sun and view. More swallows entertain us and, not to be outdone, a second-year Bald Eagle soars, not above us, but below eye level. From this vantage point, I can almost feel the lift of the breeze under its wings.

Much closer, a small bird flies into a shrub at the edge of the cliff. Binoculars up, and there, it's a Yellow-rumped Warbler, a bright spring male. This is the Audubon's race, which breeds in the west. It must be only recently back from a winter in the mountains of western Mexico. I marvel at the way its genes know what is expected of them: slate blue on the back, streaked darker; broad white edgings on the wing coverts; a black breast; and now, a bit of an accent, say, bright yellow, at the bend of the wing and, oh yes, on the crown, and an especially brilliant yellow patch there, on the rump.

The bird hops out of the bush and stands at the edge, in full splendid view on the rock and in sharp silhouette against the out-of-focus distance. The spring sun shines through air not yet sullied with summer haze, and the little Yellow-rump is stunning. Nothing can capture this: no fine-grain film, no ultra-realistic watercolour. Not yet finished with us, the warbler takes the show one step further. He throws back his head and pours out a song for all the world to hear. Then he is gone, diving into the trees downslope, with those genes urging him on to his own particular avian destiny.

It is a captivating moment, what songwriter Rick Bockner has called a "fragment of perfection." It is over as suddenly as it began, but it is more than sufficient to crown this spring morning with serendipity and magic.

Summer

American Robin

June

Barn Swallow

Grace on the Wing

The mud cup perched on a light-fixture box in the barn was scarcely cool when the female Barn Swallow settled on it again, warming the eggs of a second brood this year. Now, the mouths are again lined up inside the rim, waiting for the return of those food machines, their parents.

The Barn Swallow is probably the most widespread of the passerines (the perching birds) in the world. It is found on all continents except Antarctica, and breeds on most of them. This species is well named; it is entirely happy nesting in farm buildings, even with livestock and humans in close proximity. Barn Swallows have had access to human-built nest sites for as long as there have been human civilizations.

There was a time before barns, of course, and there are unpopulated areas where natural nest sites are still used. The swallows will nest on rock faces and cutbanks, and affix their nests to the sides of large trees. These nest sites are seen less frequently, so they aren't reported as much.

The nest, familiar to all, is built up of bill-sized wads of mud packed with bits of straw or animal hair to bind it together. It is usually plastered to a vertical surface, with a small projection to give the mud cup a little purchase. In our barn, the birds have built up against ceiling joists, but on top of metal electrical boxes. Similar locations are used in carports, with the birds showing a distinct preference for nesting over late-model, luxury automobiles.

Barn Swallows arrive last of all our swallows, so they breed later as well. The reason they are slower to arrive may be partly that their migration is among the longest of all the passerines. The species nests as far north as Alaska, and some birds winter as far south as Tierra del Fuego, at the southern tip of South America. Similar migrations are undertaken by the subspecies that nest in Europe and winter in Africa.

In the 1980s, researchers in Argentina brought to light an interesting discovery. Barn Swallows only occur there in our winter months, but in several areas they were found to be breeding, in what is the South American spring. This raises two questions: Where are they going the rest of the year? Are they returning to North America for our spring and summer? And if so, is it possible that some birds nest in South America when it is spring there, and again in North America when it is spring here?

For the rest of the North American breeders, their life history is well known. The nest is built by both sexes, and takes about one or two weeks. (This is when it's nice to have a muddy puddle or two in your yard.) The eggs are incubated almost totally by the female, a process that takes 14 or 15 days. After they hatch, the young will be fed for almost three weeks before they leave the nest and they sometimes return to the nest to roost for a few days after fledging. They are often seen perched together on fences or overhead wires, still fed by the parents for several days. Sometimes they are even fed on the wing.

Meanwhile, along the driveway, a group of recently fledged Barn Swallows chatters noisily, almost reassuringly. The first brood, and I think a few other neighbourhood broods, is pretty much on its own, and learning about life's problems. They are still dealing with nest parasites (I always knew there was a reason I didn't want to share a bed with my brother). Perched along the power line, the swallows preen constantly to remove feather mites, somehow knowing that their plumage must be in top shape to carry them to Costa Rica or South America for the winter.

There is also the matter of learning to fill one's stomach on tiny insects caught in mid-air. Having swallow genes is a bonus, but still there are some aerial miscues as the fledglings get accustomed to those new wings. Like all the swallows, they eat tremendous numbers of insects. Examination of the stomach contents of some birds has shown that they consume over a thousand mosquitoes a day! At one of the many farm reservoirs on the Saanich Peninsula, I watched a group of Barn Swallows as they foraged for insects, but some of these birds added a couple of new twists.

Many of them, and by their less-well-developed "swallow tails" I think they were immatures, were hovering near spikes of grass and the flower heads of Queen Anne's Lace. Observed through binoculars and scope, they were foraging for insects: not in flight, but on the plants. Perhaps it was just a diversion, or an opportunistic response to an insect hatch. I wondered, though, if they might have been resorting to an easier way of feeding, just until they got the hang of scooping mosquitoes out of the air.

The birds also took turns gliding below the banked sides of the reservoir and flying low over the surface. Suddenly, one plopped into the water, and then was aloft again with a shake of its wings. Other birds repeated the procedure, a new meaning to the phrase "a quick bath."

Barn Swallows leave British Columbia later in the summer than the other swallows. They stay here into September, and occasionally are still feeding young at that time. Their long flight will take them to Central or South America. Birders there do not see the long tail streamers, which are moulted. Nor do they enjoy the musical twitterings we hear from the swallows when they are here; they are much less vocal in the south.

In the following year, the birds move north again. Barn Swallows show great nest-site tenacity, returning to the same location and often the same nest year after year. Sometimes the same two birds will pair again, but they are not known to mate for life.

I am always buoyed when the Barn Swallows return. Their handsome plumage is a contrast of warm cinnamon and a cold steely blue, and their trademark forked tails are the embodiment of elegance. But I breathe a little sigh of relief, too, because Barn Swallows have gone through a sharp decline in numbers on the west coast as well as other parts of their range worldwide.

Bird populations are affected by many factors. Numbers tend to fluctuate in the natural process of matching hungry mouths with the food supply. Disease can be a factor, and swallows are particularly susceptible to nest parasites; parasitic blowflies are the main cause of nesting failure in B.C. More uncertain, and more worrisome, are the effects of pesticide use on the swallows' insect prey, not only in their North American breeding range but also in areas where the birds winter.

My impression is that we've experienced a little rebound in Barn Swallow numbers and that the population is a bit more stable. Of that I am glad, because without their grace and joie de vivre, our spring and summer skies would be barren indeed.

Mixed Feelings

In my early days as a birder, when my interest in birds was in the first stages of becoming a consuming passion, I paid a visit to the home of some friends who lived in Metchosin.

The couple I was visiting had recently seen a pair of owls in their yard, and the owls were the subject of considerable discussion. Because they had appeared in the light of early evening, my friends thought they might be Short-eared Owls, a diurnal species. Our combined references were somewhat limited, but suggested that this was not the right time of year for that species.

As if on cue, the owls flew into the yard and sat on the back fence. We examined them eagerly, noting size and field marks, confident that we could then identify them with our bird books.

Sometimes it is the minutiae that help in these identification challenges, and there was one field mark in these birds that quickly eliminated the Short-eared Owl: The eyes of these owls were dark brown. All the other marks pointed, too, at only one species; they had to be Barred Owls. The problem was, our *Birds of Canada* told us that Barred Owls were not found west of the Rockies. We could not argue with the bible of Canadian birds. I was baffled. This was the first time I had been faced with a situation wherein I was confident of my identification, but the statistics just did not back me up.

As I learned more about birds, and the birds of Victoria, it became clear that our identification had been correct; they were Barred Owls. The species has undergone a dramatic range expansion and the first record for Vancouver Island was in 1969. By the early 1980s, breeding in Victoria had been documented. Since that time, they have spread quickly over much of eastern Vancouver Island and the Lower Mainland.

This extremely adaptable species is now arguably the most commonly seen owl on the south coast. That presents a problem for beginning birders, because the owl with the highest public profile is its close cousin, the very similar — and very endangered — Spotted Owl. The two species are similar enough in appearance that it is all too easy to confuse the two. With many older bird books listing the Spotted Owl as the only one of the two that occurs in British Columbia, observers quite reasonably determine that this must be the species they have seen. It is also probably the species they have heard mentioned in news reports, so it readily comes to mind.

The reason the Spotted Owl is in the news is because it is one of the rarest birds in Canada, with fewer than 100 pairs known in this country. COSEWIC, the Committee on the Status of Endangered Wildlife in Canada, has given the Spotted Owl the status of Endangered, its highest category for a living animal. It means that the animal is in danger of extirpation or extinction as a result of the actions of humans. The entire Spotted Owl population of Canada lives in old forests in southwestern B.C. Spotted Owl habitat is also lumber-company habitat. Old-growth timber is easy to log, has a high proportion of desirable mature wood and has cost the companies nothing to develop. Species like the Spotted Owl, with no commercial value, do not figure prominently in the economic equation, so the Spotted Owl is in serious danger of losing its very home. For this reason, it has been placed on the Red List in B.C., which identifies it as a species to be monitored closely, but affords it no greater legal protection than other species.

Unlike its cousin, the Barred Owl, the Spotted Owl is not an adaptable species. Research has found that 90 percent of foraging, roosting and nesting occurs in forests at least 200 years old, and the remainder is in stands that are 70 to 140 years old. Spotted Owls also seem to have a lower tolerance for unusually warm weather, and seek out the coolest parts of the mature moist forests.

In the northwest, the diet of the Spotted Owl consists mostly of Northern Flying squirrels and other small mammals found in mature forests. Flying-squirrels are not found on Vancouver Island, and it's tempting to relate this to the absence of Spotted Owls also on the island, but at this point it has to be left as mere speculation.

Meanwhile, the Barred Owl is moving into all sorts of varied habitats, from mature woodlands to suburban parks and woodlots. It is often reported roosting in backyard trees and even fishing at goldfish ponds.

There is genuine concern that the spread of Barred Owls may be causing problems for other owl species. On the coast, their arrival has coincided with a precipitous decline in the once-common Western Screech-Owl. Barred Owls may also be pushing the less-adaptable Spotted Owls out of their habitats, and there are reports of hybridization between the two species. The fear with hybridization is that the Spotted Owl gene pool may ultimately be "swamped" by the increasing numbers of Barred Owls.

The Barred Owl would seem to have embarked on its dramatic range expansion entirely on its own steam, but on closer examination, it is almost certain that the species has benefited from human activities. Land clearing, logging and the construction of highways and hydro lines have created large expanses of the mixed-wood and edge habitats it favours. In a more direct way, the Spotted Owl is influenced by human activity; the effect in this case is decidedly negative, and threatening the survival of the species in Canada.

As we cheer the success of one species and worry about the fate of the other, we can't escape the discomfiting fact that the cause of both is a third species — us. And as much as I love to stare a Barred Owl in the face, I can't do it without a healthy dose of mixed feelings.

An update: In 2002, the population of Spotted Owls in B.C. was estimated at fewer than 50 pairs, and more recent surveys suggest that the number may be about 33 pairs. Areas occupied by Spotted Owls are identified as Long Term Activity Centres, and average about 3,200 hectares each. Under the province's Spotted Owl Management Plan, as much as one-third of this habitat may be logged. The federal Species at Risk Act passed in 2003 has provision for the federal government to intervene if it feels that a provincial government is not performing in accordance with the act but as this book goes to print, no federal action has been taken to protect the Spotted Owl. However, two logging companies, Interfor and Canfor, have voluntarily cancelled plans to log in known Spotted Owl habitat. Meanwhile, Barred Owls continue their range expansion but do not seem to have become more of a threat to other species.

Of Bicycle Tires and Daytime High Tides

Imagine, if you will, a morsel of food that is roughly as big across as you are at your waist. Imagine that it is about as thick as your neck. Imagine, too, that the outside of this tidbit is about the texture of a bicycle tire. Now imagine that you will swallow this edible delight whole. Well, the human anatomy does not work that way, of course. But this little fantasy is a fact of life for some birds.

Portland Island, north of Sidney, is a wonderful place to explore, and I often walk the trail that loops the perimeter of the island. On one visit, I stopped to enjoy the view at a gap in the trees. About 10 metres below me lay a narrow cobble beach, exposed by a summer low tide. There, an adult Glaucous-winged Gull stood patiently on a rock. I believe it must have been patient, for in its mouth was two-fifths of a Purple Sea Star. The other three rays were protruding grotesquely from the front and sides of its bill. My guess is that the sea star was about 15 centimetres across.

I have seen gulls ingesting sea stars before, but had somehow assumed that the process must be a long one: that the swallowing of the last part of the prey must follow some partial digestion of the first part. I had never committed the time I thought necessary to find out.

My attention was drawn to another gull on this little beach and it, too, had a sea star in its mouth. As I watched, it made a few gulping motions, to no apparent avail.

Meanwhile, the first gull coughed up its prey and gave it a look which could have said, "This ain't gonna work." But it picked it up again, jockeyed it into position, and partly swallowed it. The birds are nothing if not dogged, I thought.

Then, quite suddenly, the gull began swallowing again, and I watched as the sea star vanished down its esophagus. The bird's neck instantly took on the shape of its contents, with bulges here and there. The gull twisted its head this way and that, presumably trying to settle the unfortunate sea star into a reasonably comfortable position. A drink of salt water seemed to help wash it into place a little.

Not to be outdone, the second gull then proceeded to swallow its prey in the same fashion, and with the same contortions to aid the ingestion. There they stood, both with enough food to more than fill their stomachs (once it finally got there).

I was myself digesting the fact that gulls swallow sea stars in one piece. This particular species of sea star is very stiff, and rough on the outside, and I can only assume that at some point the animal relaxes to some extent so that it goes down more easily.

I wondered how long the gulls would stay here. Would they wait for this meal to digest before they moved? If they were feeding young, how long would it take for the sea star to be sufficiently digested to be regurgitated for the nestlings? Giving the kids a peanut butter sandwich began to look like a pretty good system.

To complete the performance, gull number two took a close look at a sea star left by the tide but thought better of it and backed off. Instead, it moved to the edge of the water, where it dipped its bill a few times and swished its head around as if rinsing it. It then popped its head into the water, and deftly picked another sea star from the side of the rock. It moved back to its sea-star-swallowing rock, and stood for a moment. Somehow, it manoeuvred two of the rays into its throat, and gulped hard. The sea star was now firmly in position, and would evidently take its place in line behind the first.

Confident now of the outcome, I decided to move on. One of my questions had been answered, but, as usual, others had been raised. I had seen 15-centimetre sea stars swallowed, whole, and quite quickly at that. I had seen one gull swallow one and begin on another. I didn't know if it had already eaten others. The tide had been out all morning, and there would have been ample prey.

I also marvelled at the flexibility (and toughness) of the digestive tract of the gulls. The process of swallowing a sea star looked so awkward that a tickle in my own throat lasted for some time afterwards.

But what I think I enjoyed most about this experience was the fact that these gulls were doing what comes naturally. They have been doing this trick for centuries, long before humans introduced garbage dumps and offal from fishing operations to artificially increase the available food supply.

Gulls, like all animals, must eat. On this day, the food supply was plentiful, but in winter, when the low tides occur in the darkness of night, sea stars are pretty much eliminated from the menu. This presents a major challenge to foraging gulls, and those that cannot find alternate food sources to carry them through the lean times ultimately starve.

It is the fittest that survive, to reap the harvest of the tide pools of another summer. So, while the Glaucous-winged Gull is regarded with something less than enthusiasm by many people, these two birds got a little cheer from me as I left them.

Beer as an Aid to Birding

It can be a real mental challenge to identify the more subtle sounds made by different species of birds. The Hammond's Flycatcher has a three-part song that is similar to that of the Pacific-slope Flycatcher, but the Hammond's is rougher, buzzier. In the interior, where Hammond's and Dusky Flycatchers occur together, the two are very difficult to separate. Townsend's and Black-throated Gray Warblers often have songs that seem to overlap completely. Take all of this a step further, and try to separate the call-notes of two sparrows, say.

While you are straining to differentiate these sounds, it is an aural sledgehammer that hits you when an Olive-sided Flycatcher sings from high and away. "*Quick — three beers!*" it says, with clear, whistled notes, and no mistake about it. It's one of the most distinctive songs in the bird world, and one of the easiest to learn and remember.

Now, any birder worth his or her salt knows that an Olive-sided says, "*Quick — three beers!*" Well, in the east this is certainly true, but out here, our birds shout something more like,"*I SAY there!*" with the accent on the second syllable. It isn't as much fun as the eastern version, though, and most birders continue to use beer to help them with their identifications.

In addition to being easily recognizable, the song is very far-carrying. A little of it goes a long way, in the sense that a single bird can be heard over a wide area. What is a little worrying, though, is that the total number of birds can be relatively few.

The Olive-sided Flycatcher is one of the most widespread breeders across Canada, and is found almost everywhere in British Columbia. It likes mature coniferous forests for the most part, but needs openings in which to forage. These can be natural openings around ponds or burns, and open hilltops are favoured spots. Openings left by logging or other human activity may also provide habitat. The big flycatchers are almost always seen atop a tall snag, flying out to catch insects in mid-air and then returning to the same perch again and again.

They sit erect, as flycatchers do, looking a little as though they are puffing out their chests, and their white shirts show under their dark vests. Fluffy white feathers on their flanks sometimes protrude from behind their folded wingtips. Exposed against the sky, they can be seen from almost as far away as they can be heard.

While they may seem to be widespread, North American Breeding Bird Survey data from the 1960s through the 1980s have shown that Olive-sided Flycatcher numbers are falling off sharply, with the most serious declines in B.C.

One suspected cause is the deforestation of the birds' neotropical wintering areas, mostly in Venezuela and Colombia. This is a very real problem for many North American species. There is also a suggestion, though, that the decline may be caused by loss of breeding habitat to logging in North America, and the concomitant loss of prominent snags from which to forage. The subsequent replacement of the habitat with even-age monoculture forests would be unconducive to Olive-sided Flycatcher nesting.

It is possible that the survey data are indicating population changes resulting from habitat changes along the survey routes. Absolute numbers may not have changed; rather, the birds may have moved to more suitable habitat as it became available. That may also be wishful thinking.

The Olive-sided Flycatcher is rarely a host to Brown-headed Cowbird eggs, and this may be because the flycatcher prefers to nest well into the forest. Cowbirds seldom venture farther than 300 or 400 metres in from the forest edge. Low brood parasitism is the good news; the bad news is that Olive-sided Flycatchers are dependent on large tracts of mature forest, which are disappearing at an alarming rate.

If Olive-sided Flycatchers are worried about this, they aren't letting on, and there is something very positive in that. While some birds go about their lives in furtive seclusion, these guys let the whole world know that they're here, they mean business, they're not leaving until the fall, and they don't have time for whiners and complainers. They have, as the saying goes, an attitude. Whether you are a beer drinker or not, you have to admire a bird that looks at life that way.

American Dipper

July

A Clear and Present Danger

It arrived just as so many others have in the past, neatly wrapped in tissue. Partly it is out of respect, I think, but also there is a conscious effort not to damage a thing of such beauty.

It was not brightly coloured; in fact, it would be difficult to describe a plainer bird than this. Greenish-grey on the back and creamy white underneath, it had no dashing wingbars, no contrasting streaks, no splash of colour on rump or crown. Its only mark of any consequence was a faint white line above its dark eye. But every one of its plain feathers was a study in perfection, each intricately constructed to produce a member of great structural strength and tremendous versatility.

The feathers of the body provided aerodynamic shaping, but they would also have been able to expand greatly, trapping insulating air close to the skin on a damp winter morning in the upland forests of El Salvador. Tiny plumes arranged themselves exquisitely to cover the head and face completely, leaving only those important eyes fully exposed. Stiffer feathers lined the wings and tail, incredibly light but strong enough to carry the owner across thousands of miles of skies, against wind and rain and drying sun.

There would have been times when these feathers effected an escape from the stoop of a predatory falcon. In this case, they were not enough to bring the bird to a halt before it killed itself on the blue sky-turned-barrier we call a window.

The bill of the bird was a giveaway. Uniformly thick through its length, it had a small but distinct hook at the tip; it's a classic mark for the vireo family. The plainest vireo in North America is the Warbling Vireo, and this bird was one of those that had returned from the neotropical winter, to nest in British Columbia.

I examined the bird's belly and, in parting the feathers there, exposed a patch of bare skin. It is called a brood patch, and it allows the body warmth of the bird to incubate the eggs, without fighting the insulating effects of those marvellous feathers. We might assume that this was a female, but in vireos, the sexes are similar in appearance, and both share in the incubation so, without a bird bander's reference guide, I could not be sure. Either way, the bird was in breeding condition.

Tremendous numbers of birds are lost to windows. While many perish in collisions with tall artificial structures, often the most visible losses are those in suburban backyards. The most effective way to limit these deaths is to remove the threats — large expanses of glass. Of course we cannot live without windows, but they need not be as large as many are. The smaller windows of older houses, divided into many panes, each reflecting light at a slightly different angle, present more broken reflections, and birds seem to collide with them much less frequently. Guilty also in the avian slaughter are glass deck railings, which certainly don't fall into the category of indispensable.

Another solution is to prevent the birds from hitting the glass. The most effective way is to stretch fine garden netting outside the window, so the birds bounce off harmlessly. Falcon silhouettes may work a little bit in some places some of the time (how's that for an endorsement?) but will be most effective if free-swinging and falcon-coloured. Some people have found that hanging Japanese-style wind socks in front of the windows reduces collisions. In rooms with glass at either end, closing curtains will help prevent birds from trying to fly right on through.

If you do find a bird that has stunned itself in a collision, first determine if it is safe from predators where it is — if so, leave it. If not, place it in a shoebox or other dark place to recover. It may take a matter of hours before you hear the scrambling that indicates that the bird is feeling better and can be released. In most cases, birds will either recover from the shock, or will die of brain injuries. Birds that are injured should be taken to the nearest wildlife rehabilitation facility; the SPCA or the provincial conservation officer can point you in the right direction.

And what about those beautiful feathered corpses? With the exception of a few "nuisance" species, it is technically illegal to keep birds or their parts, including feathers. They can, however, be used in a variety of ways. The Royal B.C. Museum will accept birds, but because of limited resources may keep only unusual species. The biology department at your nearest university or college can also use the specimens, or may direct you to other researchers who have permits to use them. Sometimes, institutions in different parts of the country exchange species common to their areas, to expand their collections.

It is poor consolation for the death of a Warbling Vireo, but if we can learn something as a result, there is at least some small benefit, a way perhaps to mitigate the declines these birds suffer as a result of our activities in their world.

The Revenge of the Juveniles

At this time of year, a whole new contingent of birds arrives on the scene. They're not migrants or vagrants or truants. What makes them different is just that they are very young. We should expect this. This is the season when the territory marking and courtship displays and long days of feeding gaping mouths have produced fruit. The next generation of birds, the promise of the future, has fledged and is on the wing.

This is also the time when the wonder of reproduction takes a decidedly pernicious turn, at least as far as birders are concerned. It was just getting manageable, this game of identifying birds. The males in their breeding plumage looked just like they did in the field guide. With the males under control, it was possible to tackle the next challenge.

Even some of the females could be identified. Some were distinctive enough to be identified by sight. Others obliged by sticking close to an identifiable male. But then, the honeymoon ended. It wasn't a sudden thing; it started with an unidentifiable little brown bird. But then there was another, and two more, and that little swell of confidence eroded away. It happens every summer; it is the Revenge of the Juveniles.

They are everywhere. They don't look like adults. They don't sound like adults. And they don't behave like adults. What's worse, for every pair of adult birds, there are anywhere from 2 to 12 unidentifiable juveniles that have entered the game, and they're playing for the other team. But don't despair; there is hope for birders who are trying to sort out the summer onslaught.

Several clues will help in the challenge of identifying juvenile birds. The first task is to determine if the bird is in fact a fledgling, rather than some tantalizing new species.

Young birds move in a distinctive way. They are usually a little clumsy, venturing short distances in a flight that is characterized by rapid wingbeats and slow progress. Landings are also a matter of little experience and considerable experimentation. They often involve acrobatics to stay on a perch, or noisy crashes into the understorey. It makes perfect sense that these are young birds, still growing into their new bodies.

Juvenile birds usually appear unkempt. They often have a fluffy look about them, and may have bits of down still showing in their plumage. Young House Finches have two distinctive horn-like feathers on their crowns for a time. Another clue to the age of a bird is its mouth. Young birds have less feathering around the gape, or edges of the mouth. Usually the bare skin at the gape is a quite bright yellow, and is visible at some distance.

A dead giveaway is a bird that is being fed by another bird, one that appears to be an adult. This is a great opportunity to learn what the young of a species looks like, because the identifiable adult is right there for reference. Beware, though, of Brown-headed Cowbird young, which are always fed by a surrogate parent; usually, the hard-working adult is much smaller than its foster young.

Often it is a sound that draws attention to a bird, and young birds introduce a whole new phalanx of vocalizations to confuse birders. It can be difficult to identify young birds by their calls, but it pays just to remember that unfamiliar sounds may be coming from young birds. If you can find the source, you may be able to identify it visually.

Once a bird has been aged as a juvenile, is it possible to identify it as to species? The answer is an unqualified "sometimes." There are birds, like herons or kingfishers, which are difficult to mistake. And then there are sparrows.

Many of the smaller birds begin life in a streaky plumage. The body feathers will, in most species, be moulted soon but often, the flight feathers of the wings and tail are kept, and look the same as those worn by adult birds.

You may notice, scratching in the leaf litter, a bird about the size of a towhee, heavily streaked with brown. If it flies, however, you'll see the familiar flash of white spots in the wings and tail; it is a towhee, but a bird not long out of the nest. Another streaky brown job is the juvenile junco. It doesn't look at all like its parents, until it flies. Then, with a flash of white in the outer tail feathers, it becomes a Dark-eyed Junco.

Some field marks are valid regardless of the age of the bird, like size, or the shape of the bill. (It's pretty difficult to misidentify a juvenile hummingbird.) But other marks change as birds mature. Young Cooper's Hawks have brown eyes, which become yellow by their first winter and red as they reach adulthood. Barn Swallows have very long forked tails, but fledglings aren't so well endowed and may even be misidentified.

It's a time of confusion, all right, but once you have identified a few juvenile birds, it starts to get easier. As you regain a little of that eroded confidence, you'll begin to see the Revenge of the Juveniles for what it is: just another birding challenge.

The Potato Chips of Summer

The lush green of the fields is gone. The stillness of the summer air is broken only by the rustle of dry grasses against each other. The vitality of spring seems to have vanished. Then, across the field comes a posse of small birds. They bound up and down as though on a roller coaster, and at every rise they call out a cheery *per-chick-o-ree*! (Some hear the call as *po-ta-to-chips*!) Their ebullience is matched by their appearance: bright lemon-yellow bodies with black berets, powered by black wings with flashing white wingbars.

Their flight is not random, though. They arrive, putting up in a drift of thistles that has survived along the edge of the field. Suddenly, there is a reason to appreciate thistles. Among the ripening flower heads, these little yellow birds cling with tiny feet, delicately extracting seeds.

They are American Goldfinches. Wild Canaries. Thistle Birds. There are few birds that are so distinctly tied to one plant as the goldfinches are to thistles. They relish the seeds, although they also eat the seeds of other plants, and they use the thistledown in the construction of their nests.

Goldfinches are so connected to the life cycle of this one plant that they nest late in the season, when at least some of the wild thistle crop has ripened. It's not unusual for them to wait until June or July to start raising a family. The nest is remarkably well made. It is neatly woven of grasses, and lined with thistledown or sometimes, in the east, milkweed down. It is so well assembled that it holds water. It's usually found in a hedgerow or open wooded area, most often about one to 10 metres off the ground.

Birds of open country are very vulnerable to nest parasitism by Brown-headed Cowbirds, and American Goldfinches are in exactly that category. Some sources report that goldfinches are frequent victims of cowbirds but there is one reference which claims that the goldfinch's late nesting does not coincide with the earlier nesting urge of the cowbird, and as a result cowbird success is minimal. The hypothesis makes good sense, and the subject would make an interesting study.

As these birds feed noisily in the weeds, you can see that some look different, less brightly coloured. The females are more subdued, with greenish-yellow bodies, but they still have dark wings with prominent wingbars. The bill of both sexes is pale, often pinkish.

If they are feeding young, these birds will collect seeds, swallow them and return to the nest. There, they regurgitate a porridge of partially digested seeds. This behaviour is unique to the finches; other seed-eating birds feed protein to their young in the form of insects and other animal matter.

Their year-round diet of seeds means that goldfinches (and other finches) are readily attracted to feeders. They relish black oil sunflower seed. Goldfinches in particular are fond of niger seed (which is why it is often sold as "thistle"). If you decide to offer it to them, use a feeder designed to limit the flow, or a special "thistle" sack, to make this expensive seed go as far as possible.

As seed eaters, goldfinches provide a great service in eating huge numbers of weed seeds. They can often be attracted to the garden to feed on the seed heads of lettuce, cosmos, coreopsis and other plants — a good reason to hold off on deadheading them. A few insects and berries are also eaten, but it is the thistles to which the goldfinches are most loyal.

As the summer wanes, the yellow of the males begins to be replaced by duller green for the winter. Many goldfinches will leave southern Canada for open country in the southern United States. A few usually stay to winter on southern Vancouver Island, but it is not every feeding-station operator who will have them around to enjoy.

They often appear in loose company with their close cousins, the Pine Siskins, and sometimes with House Finches. One of the things that have struck me about these goldfinches is how small they are. I think of them as being as big as House Finches, but they are really only about the size of the chickadees.

Perhaps size has something to do with their apparent zest for life. Like the chickadees, nothing seems to faze them; they just move on to something else. They add a little bubbly chatter for good measure and when they leave, they do it in roller-coaster style.

I've cut a lot of thistles on the property, but there are a few that I ignore. It's not because I think they'll go away on their own. Rather, I'm afraid that the goldfinches might go away. So a few thistles will stay and, with luck, a few goldfinches, too.

The Seabird of the Old Forest

The kilometres rolled by beneath the wheels of the little boat trailer, courtesy of the new bypass routes of the Island Highway on Vancouver Island. The final leg along the old route into Courtenay was short, and brought my family and me through the town of Comox and down to the ferry terminal at Little River.

This ferry, less well known than others farther south, skirts the north end of Texada Island and docks in the town of Powell River on the Sunshine Coast. We drove off the ferry and headed north, past the stately wooden houses in the older part of town. Our destination was a bed and breakfast near Okeover Arm, which would be our base for several days' exploring in the area around Desolation Sound.

Although Desolation Sound is well known to cruising families, I hadn't realized how accessible it was in a small boat. Our rigid-hull inflatable hurried us along in the protected waters, allowing us to explore many of the spectacular bays and islands in the marine parks. The boat was also small enough that we could sneak into some of the less accessible beauty spots, too.

Desolation Sound was named by an inexplicably depressed Captain George Vancouver. On a gloomy day, the steep-sided inlets loom powerful and dark, it's true, but Vancouver would have seen them in their splendour, too. His unhappiness seems to linger in Desolation Sound, for with each new discovery we made, his name came to mind.

We tied our boat to the dock below Cassell Lake, and walked the short, steep trail to the lake. A male Yellow-rumped Warbler was engaged full-time in a Lodgepole Pine, feeding a streaky juvenile bird. Beneath them, the short outflow stream tumbled toward Cassell Falls, which fell, with delicious serendipity, over the lip and directly into the sea below. Vancouver's men swam and bathed in the lake, but there is no record, apparently, that any of them saw Yellow-rumped Warblers. Vancouver did not put much emphasis on natural history, on this voyage, at least; he just wanted to leave.

Neither is there any record of the men seeing what we found to be the most common bird in these waters: the Marbled Murrelet. They would have been here, and more numerous than now. They are still quite common, though they are considered threatened.

The Marbled Murrelet story is one of mystery. The bird is a member of the alcid family, most of which nest in burrows or on the ground. Where do these murrelets nest? For years, nobody knew, at least not for certain. Then, in the last few years, speculation was confirmed. They nest in old coniferous trees, as much as 50 metres off the ground and as much as 50 kilometres inland.

This species, like the Spotted Owl, is dependent on extensive stands of old-growth timber for its nests. After a month or so in the nest, the single young bird must leap from the large, moss-covered branch it has called home and fly successfully to the sea.

It's estimated that the murrelet population has declined by as much as 40 percent because of loss of its old-growth habitat to logging. In an effort to determine how serious the problem is, researchers are attempting to learn everything they can about Marbled Murrelets. One important research tool is to band the birds. Murrelets, however, are almost always seen in flight (fast and overhead) or on the water (or diving under it). Leg bands just aren't visible. The solution is coloured wing tags. First, you have to catch them.

Many birds are caught in mist nets set across flight routes. Murrelets spend much of their time on water, which is a tricky place to set up a mist net. It's been done, though, and people have also had good success using dip nets to capture the birds as they feed at night. Much has already been learned in only a few short years, and the work will continue.

Still, the mystery lingers. There is little historical information about these tree-nesting seabirds. Their status is still uncertain: Not long ago, the Asian race was deemed to be a species entirely separate from the North American breeding population, and was given the name Long-billed Murrelet.

One thing is known for certain. The Marbled Murrelet is entirely dependent on the mature coastal forest ecosystem for its survival, and there will need to be more protected areas along the coast to ensure a future for the species. Ironically, as humans have protected areas around Desolation Sound from logging for the aesthetic pleasure of summer boaters, we may have given the Marbled Murrelet at least one place where it is secure for a time.

Nighthawk

Nighthawk. It could be the name of a fighter aircraft. Perhaps an express train, moving business people from meeting to meeting while others are sleeping. It might even be a comic-book character. In truth, nighthawk is the name of a bird. Well, you say, this sounds like a pretty interesting bird. But if we go one step further, we find that the name itself is at once apt and misleading.

In Canada, the nighthawk we can expect to see is the Common Nighthawk. And yes, we can expect to see it at night. We can also expect to see it during the day. Most of all, we can expect to see it in the evening, when the day has not yet yielded dominance to the night.

What we cannot expect to see is a hawk, for this the nighthawk is not.

When seen at rest, the nighthawk typically sits, rather than perches, because its feet and legs are quite weak — not the fearsome talons of a Merlin. These birds also tend to align themselves lengthwise along a branch or fence rail, and their cryptic colouration helps them to blend in with their chosen perch — not the alert posture of a Red-tail, scanning its territory from a high and visible perch.

The dissimilarity does not end there. The business end of a true hawk carries a prominent, sharply hooked bill, for dispatching and eating prey. The nighthawk, in contrast, has a tiny, flat bill. What is remarkable, however, is that the lower mandible changes shape to create a very large gape. The nighthawk feeds by flying with its wide mouth open, scooping up insects as it goes. A hunter, yes, but far from the image of the stooping Peregrine.

Why, then, is it called a nighthawk? The first part of the name is at least somewhat accurate — the birds are most active at dusk. But it is not uncommon to see them at midday, particularly in areas where they may be nesting.

The "hawk" part of the name refers to the bird's flight silhouette. The wings are long and slim, swept back at the wrist, and taper to a pointed tip. Easily visible as they pass overhead are the prominent white bands across the wingtips. These long, pointed wings, hawklike in their resemblance to the wings of a falcon, are flapped, but in a very un-hawk-like fashion, rather slowly, up and down, through a deep arc. Yet the bird moves beautifully through the evening sky.

This is British Columbia's most widespread member of the bird family known as the nightjars. Elsewhere, its cousins say their names; "*Poor-will*" in the Okanagan, "*Whip-poor-will*" in eastern North America, and (things go from bad to worse for Will), "*Chuck-will's widow*" in the southeastern United States. But our country cousin, the Common Nighthawk, says only *peent*. It is not musical, not ethereal, not bone-chilling, not anything special, and certainly not name-saying. It's just a nasal sneeze of a call.

The male does say it often, though. It is an important part of his approach to a mate. With such a limited repertoire, nighthawks have also evolved another method of communicating their ardour to their intended. Their courtship flight ends in a steep dive, the male levelling out very close to the passive female on the ground. As he pulls up, he bends his wings forward, and the wind rushing through his primaries adds a buzzy bass *voo-o-o-m* sound to the lovers' concerto. It is more than a little disconcerting to hear this and have no idea where it is coming from.

One lazy July day, I visited Cordova Spit, one of the few sand-dune habitats on southern Vancouver Island. My route was erratic as I sidestepped clumps of Dunegrass and avoided trampling the Beach Morning-glory. Then, in front of me, the sand exploded, and a nighthawk flopped away on long, swept-back wings.

I pinpointed as nearly as I could the precise location where this bird had been. With great care, I walked toward the spot and, after a few minutes of searching, I found its nest. I use the term loosely; what I found was two eggs, heavily splotched and extremely well camouflaged, lying side by side amid some woody debris on the sandy ground. I left quickly, and watched as the adult bird returned to the nest.

On Vancouver Island, Common Nighthawks also take very nicely to regenerating clearcuts for nesting — some pairs return to the same old logging roads for two and three years. But clearcuts (thank the gods) are not static, and the birds eventually must move. In urban areas, nighthawks are declining, because of the loss of available open areas for nesting. There are recorded nestings on gravelled roofs, but this is not a preferred situation.

The nighthawks will soon be more visible overhead, as adults forage to feed hungry juveniles, and the juveniles themselves take wing to fend for themselves. When flying ants and termites leave their colonies, the nighthawks are there to take full advantage, in a spectacular show of aerobatics.

Nighthawks are well equipped for their niche. Their distinctive wings give them admirable aerial ability. Over 2,000 flying ants have been found in the stomach of a nighthawk, showing the efficiency of their feeding system.

Large flocks of nighthawks begin to form prior to their southbound migration, and may suddenly appear in the sky on a late summer evening. Their migration takes them to South America, some as far as Argentina. They travel not so much to a different continent as to a different summer, where their diet of insects can be satisfied.

There are warm summer evenings ahead, punctuated by the sneezing of nighthawks. They will become less vocal as the breeding season winds down, but still they will be foraging overhead. They grow on you, and then, all of a sudden, they're gone. But not to worry; they'll be back, and better still, they'll bring the summer with them.

Pileated Woodpecker

August

Baja British Columbia

There is an island in the Gulf of California called Raza Island which has little to recommend it except that it is the breeding ground for the majority of the world's Heermann's Gulls. Some 600,000 pairs of the elegant grey birds nest there every year, with the rest scattered elsewhere along the gulf's Mexican shores, and on the Pacific coast of Baja California.

What is of interest to British Columbia birders is that, when the breeding season finishes in June, Heermann's Gulls disperse in all directions. Many move farther south along the Mexican coast, but the majority moves northward, with good numbers reaching the Strait of Georgia, starting in early July.

It's always a pleasure to see these gulls return because they are among the most beautiful in North America. The adults are a velvety slate-grey above, shading to black on their primaries and tails, and paler grey underneath. The trailing edges of wings and tail are pure white, as is the head, and the picture is finished by the addition of a bright red bill, tipped with black. They are a treat for the eye, and easily distinguishable from all our other gulls.

In all gulls, the adult plumage is not attained in the first year; the time varies, according to the size of the species, from two to four years. Heermann's take three to complete the job.

Heermann's Gull is almost strictly a coastal and marine species. Victoria is a good place to see them, just offshore at places like Race Rocks; they are especially fond of resting on floating kelp, logs or other debris. But they do come to shore, and can be seen in many places along our coast, usually in the company of other gulls.

They often appear in the mixed feeding flocks of seabirds that can sometimes be seen close to shore. The first birds to arrive are the Glaucous-winged Gulls, thrashing around and attracting attention. A few Pelagic Cormorants come in, diving for food as they do, and avoiding the marauding gulls as much as possible. Along the edges, Bonaparte's Gulls delicately pick bits off the surface, and Rhinoceros Auklets make repeated dives. The Heermann's Gulls fall somewhere in the middle in terms of size, and use a variety of feeding tactics, gleaning bits that have been missed and pirating whatever they can.

Heermann's Gull was first reported to the world of western science about 1850 by, you guessed it, a man named Heermann: Adolphus Lewis Heermann. Heermann was a United States Army surgeon who, as was often the case, was also a dedicated naturalist. Born about 1827, he had graduated from medical school by the time he was 22 and was in the west, beginning a large collection of bird skins. He was involved in the Pacific Railroad Surveys for several years, and remained in the west during this time, working closely with another influential American ornithologist, Spencer Fullerton Baird.

Apparently his health forced his retirement to a cattle ranch in Texas when he was only 36. It was probably an advanced case of syphilis that brought this on; in his last years he was almost unable to walk unassisted because of the disease's effect on his nervous system. He continued to collect until the end, however. Two years later, when he was 38, while in the field, he accidentally shot himself, and did not survive the injury.

The gulls that carry his name are here to remind us of him. They will be with us for a few months yet, biding their time along our shores. The birds begin to leave in September, and slowly work their way back to the Gulf of California. They are usually all gone by the end of October; winter records in our area are rare. On a field trip to Rocky Point, west of Victoria, in March of 1986, one was found all by itself on a dock. It was a second-year bird, and probably had not felt quite the same urge to move south as other Heermann's Gulls of breeding age. Whatever its reasons, it brightened our day because we weren't expecting it.

I reversed the roles a bit myself, when I visited the Heermann's Gulls in their home territory in Baja California. The scene was familiar, with a variety of seabirds converging on a small patch of ocean as a feeding flock developed. The difference was in the cast of characters.

The Heermann's Gulls were edged out of the way by the larger Yellow-footed Gulls, but they held their own with the black-hooded Laughing Gulls. Royal Terns, large and graceful, patrolled the edges of the mêlée. Brown Pelicans plunged their gangly shapes into the sea in vertical dives, and Blue-footed Boobies joined the fray as well.

In the warmth of a Baja winter, the Heermann's Gulls again establish their territories on Raza Island and the other breeding colonies. When the work is done, they will begin to disperse, heading north to our waters. It's a different sort of seasonal movement from what we normally expect in birds, and it adds an exotic touch of the tropics to our temperate summer.

A Shorebird Identification Clinic

It's traditional about this time of year for bird writers to soothe the fevered brows of anguished birders caught in the horror of the shorebird migration with inadequate identification skills. The thought of the many different species in their varying plumages can be daunting, but a simple system will help to sort out even the most challenging of these shore-loving nomads.

Let's start with an easy one. Here's a bird with long, yellow legs. A Greater Yellowlegs. It's a yellowlegs because it has yellow legs. It's a Greater Yellowlegs because there is another bird with yellow legs that isn't quite the same. Well, it looks the same, but if you put it beside the first one, it's smaller. So it's called a Lesser Yellowlegs. Anytime you see two birds together with yellow legs, then, you'll know for sure you have a Greater and a Lesser Yellowlegs.

The Terek Sandpiper has only been seen once in British Columbia. It has yellow legs, but it is not a yellowlegs. Identification must be done carefully, but remember that if you do not see one very often, it could be a Terek Sandpiper.

Consider also the Wandering Tattler. It is so named because it migrates

— a useful clue. Its Asian cousin also migrates, but because "Wandering" Tattler was already taken, it was named the Gray-tailed Tattler. Both of these birds have grey tails and yellow legs.

A similar situation exists in the elegant godwit family. North America's Marbled Godwit has a marbled plumage and a barred tail, as does Asia's Bar-tailed Godwit, and it was simply a matter of who got in the name lineup first. It is possible to separate these two species, however, from the Black-tailed Godwit, which has a black tail, as does the Hudsonian Godwit.

One of the most misunderstood identification challenges is the family of small sandpipers known as "peeps." The name mimics their soft calls. In Britain, the birds have a different accent, so they are called "stints."

The Western Sandpiper is the most common migrant in the eastern Pacific. It has black legs, when they are not muddy, and bits of webbing between the toes. For this reason, Westerns might have been known as "Semipalmated" Sandpipers but that name was used up first in the western Atlantic region, by the Semipalmated Sandpiper, which has tiny webs between its toes, too. It may help to note which ocean is nearest you.

Size can be a useful aid to identification. One of the smallest of these sandpipers is called the Little Stint. It is a Eurasian species and is thus unlikely to be encountered on our west coast, but these are long-distance migrants and should not be discounted entirely. The Little Stint is little, only slightly larger than the Long-toed Stint. Even smaller is the North American Least Sandpiper. Its yellow legs will distinguish it from any vagrant Little Stints. Long-toed Stints also have yellow legs, but of course can be separated from the others by the length of their toes, which they use to disturb the mud in the shallow water in which they habitually stand.

The plover family presents its own challenges and their attendant solutions. The species most familiar to us is the only one with two black bands across its breast, so it is called a Killdeer. If you see a very small plover with only a single breastband, it could be a juvenile Killdeer. Check the feet, though, for if they are slightly webbed, it is a Semipalmated Plover. If not, it could be a Common Ringed Plover. If it is one inch smaller, you have a Little Ringed Plover. You may need binoculars when working with these species.

You are now ready to tackle the dowitchers. These birds use a feeding motion similar to a sewing machine's action; the name replicates the machine's sound. The Long-billed Dowitcher has a long bill, as does the Short-billed Dowitcher, so if you see a bird with a long bill, it may well be one of these. The surest way to separate them, though, is to ask.

There you have it. You need never be daunted again by a flock of mixed shorebirds. Simply remember that you are not the first to have faced these problems, and be thankful for the efforts of those who have walked these shores before you; they helped to smooth the bumps on the challenging road to shorebird identification.

The Loss of a Giant

Birders often measure their success with lists: how many birds they have seen in Canada, in British Columbia, Victoria, their own backyard; how many birds photographed; how many identified by sound alone; how many recorded on audiotape. Some list how many new species they have discovered.

There are few, of course, who can make this last claim, of the discovery of a species new to science, but it still happens, almost every year. Most often it is in South America, sometimes Southeast Asia. The last new species reported in Europe was in 1975.

In North America, birders strive for landmarks on the lists of birds they have seen. Three hundred and 400 are easy, 500 and 600 progressively more difficult. There are not many who have recorded 700 or more species in North America. This represents about 80 percent of the number of species known to have occurred in North America, and it is a remarkable achievement, requiring an extensive knowledge of field marks, behaviour and habitat preferences.

Consider that in Peru — a nation only a third larger than B.C. — there are about 1,700 species. Consider, too, the difficulties in birding in such a country, with poor roads, no field guides, and little English spoken. Imagine a birder who has seen 1,675 of those species. Ted Parker accomplished that.

His name is hardly a household word. He was not well known to the average birder but in the world of neotropical ornithology, he was a giant. Parker, possibly the most superb field ornithologist the world has ever known, was killed on August 3rd, 1993, in the crash of an airplane in which he was conducting low-level forest surveys in Ecuador.

Ted Parker had an early interest in birds as he grew up and, while still in high school, he broke the Big Year record for the United States. His college years took him to Arizona and Louisiana, where he was introduced to birds of the neotropics. As he settled into a career in biology, his research soon overtook his interest in listing, and he spent much of his time doing field studies in the neotropics.

His abilities were truly amazing. In the multi-level canopies of tropical forests, birding by sight is a very limiting pursuit; you have to learn to use your ears. Parker's hearing was legendary. He knew the songs of a tropical family of birds called antthrushes well. On one occasion, he heard a song that was unfamiliar to him, but he thought must be an antthrush. He identified it as that of a Rufous-fronted Antthrush, known then only from two specimens taken 30 years earlier.

He could identify mystery birds on tape, and would also identify the songs in the background. On one recording, he picked out an undescribed species of antwren, based on his knowledge of the songs of the known species. Research by another scientist proved Parker's identification within a year. Parker discovered, again by sound, one or perhaps two new species of birds in Ecuador, just days before his death. In addition, he was an expert in the recording of bird songs in the field. Of 90,000 recordings in the Cornell Laboratory of Ornithology, more than 10,000 are Parker's.

His research was not limited to the identification of birds: He knew his birds so well because he studied them. His understanding of bird behaviour and habitat preferences was remarkable, and he quickly learned that in South America they were as important field clues as visual and auditory ones. Parker's work in South America was described in a 1990 book entitled *A Parrot Without a Name* by Don Stap; it is a fascinating look at the world of neotropical bird research.

Ted Parker did not limit his efforts to research. He also spent time leading tours, and his success at finding birds, and getting them close enough for clumsy North Americans to see them, was well known. In later years, he was turning more and more to conservation work. He knew, better than anybody, that it was a race against time to save as many species as possible. And he knew, perhaps more than anybody, that some of those species did not even have scientific names yet.

I cannot tell you anything about the man personally — his home life, his loves. Perhaps he was one of those so driven by his work that it became who he was. Certainly he was immersed in his work, and set examples that few people will be able to match.

Ted Parker was unique, and we should not expect that we can be like him or feel that our accomplishments and those of others are diminished in his shadow. The work he did in his 40-year life will be continued by others but still, when such a giant is so suddenly and irrevocably gone, one can only wonder what might have been.

In the Geological Wink of an Eye

On August 27th, 1883, the world literally shook, as the volcanic island of Krakatau was disembowelled by a series of violent eruptions. The island (sometimes called Krakatoa) lies between Sumatra and Java, in the Sunda Strait. The explosions, heard as far as 4,600 kilometres away, reduced Krakatau to a third of its former size. The remaining mountain, called Rakata, was covered with 40 metres or more of pumice and obsidian, at temperatures high enough in some places to melt lead.

The cataclysmic and complete end of all life on Krakatau provided an exceptional opportunity for scientists to watch as the isolated island ecosystem regenerated over the years. As early as nine months after the event, a French expedition, almost skunked, discovered one tiny spider, the first species known to recolonize the new Krakatau. Other species followed, many windborne, some by swimming or in flight, or aboard ad hoc rafts of various floating materials. By 1886, there were already 15 species of grasses and shrubs, and by 1928 nearly 300. A large monitor lizard was reported in 1899.

Intensive studies in the mid-1980s revealed as many as 30 species of land birds and nine bats, among an ark-full of other species. However, the present flora and fauna are not the culmination of a process uninterrupted since 1883. Many more species found their way to the island but a lot of them did not survive. Some undoubtedly became established, only to be extirpated some time later. The whole process has enabled scientists to reach a better understanding of how island ecosystems evolve and operate.

The eminent biologist Edward O. Wilson and a colleague, Robert MacArthur, together developed a theory of island biogeography, and their thinking was certainly influenced by the Krakatau experience. Their theory states that as islands increase in size, the species diversity increases as well. They found that for every tenfold increase in land area, the species number will double. Wilson and MacArthur were quick to point out that there are other factors at work as well. It is obvious that if an island lies close to a larger land mass, its species diversity will be greater than if it is isolated by thousands of kilometres. Even so, in a group of islands equidistant from a larger shore, the larger ones will support more species.

It seems an obvious principle and it has a much greater impact when put into a local perspective. There are more species of birds, for example, on Vancouver Island than there are on Salt Spring Island. And there are more on Salt Spring than on Wallace Island, off its northeast shore. Take it to its extreme, and you will find only a few species of land birds breeding on some of the austere little islets near Wallace.

The Wilson-MacArthur theory also applies to other islands, those isolated not by water but by expanses of very different habitat. A deciduous copse in the middle of a prairie will host many fewer species than a larger forest, for example. This presents a very compelling argument for sober second thought about the reduction of wildlife habitat below sizes that will support a healthy species mix.

Krakatau has now regenerated to the point where it looks much like any other Indonesian island, but 121 years is nowhere near enough for biodiversity to maximize its potential on the island. And in the end, the result will almost certainly not resemble the careful mix of species that had sorted itself out in the conditions that existed prior to 1883.

It has given us an opportunity to compare theory with reality, and then back again. In a geological wink of an eye, we have been privy to a process as old as the Earth itself. Best of all, people like Wilson and MacArthur are forcing us to actually learn something in the process. There may be hope for our species yet.

Another Heaven and Another Earth Must Pass

The early ornithologist Alexander Wilson watched as the birds passed overhead. The flock increased in width until it covered three miles of sky. And on the birds came, darkening the sky, a continuous, streaming river of birds, over three hundred miles from start to finish. Wilson estimated that there were over two billion birds in the flock.

They would gather in huge colonies in the hardwood forests of eastern North America. Millions of birds would occupy an area of several hundred square kilometres, with as many as a hundred nests in any tall tree. This was the Passenger Pigeon, probably the most abundant bird ever to live on the planet.

In the southeastern United States, a unique little parrot would gather with 10 or 20 others of its kind in a tree cavity at night, to roost. By day, they would form larger flocks, flying erratically over the forests, screaming and chattering constantly. The Carolina Parakeet lived farther north than any other member of its family, and in huge numbers, too.

The Native people harvested young Passenger Pigeons that had fallen from the nest trees, and they found more than enough to satisfy their needs. The Europeans quickly learned to use this resource, too, but they soon began to employ much more efficient harvesting methods, as the pigeons became a commercial commodity. Using bait, and live birds as decoys, trappers netted the birds by the thousands. In one month, from one colony in Michigan, over 700,000 birds were shipped to market. As the railways expanded westward, the trappers could tap the colonies in new areas. Thousands of birds were used as targets in shooting galleries.

At restaurants in the east, diners raved about the succulent squabs. It was the fashion for women to wear elaborate hats, and the hats of many of the restaurant patrons were decorated with the bright green feathers of the Carolina Parakeet.

If the parakeets evaded the plumage hunters, they were shot by farmers, who destroyed the gregarious birds in huge numbers because they damaged fruit and grain crops. The parakeets made the job brutally easy: They would gather where one had fallen, so that entire flocks were ultimately destroyed. Many were also taken for the cage-bird trade. The birds that survived these depredations found that their habitats had been severely depleted in the early rush by settlers to colonize new territories.

By 1880, several states had become alarmed by the decline of the Passenger Pigeon, but legislation to protect the birds was rarely enforced. At the end of the decade, the decline was almost complete, and the last bird was seen in the wild in 1900. A year later, the last specimen of a Carolina Parakeet was taken, and the last bird was seen in the wild in 1904 (though other controversial reports exist).

Both species continued to exist in captivity. The parakeets, though, were inattentive parents and not many young were successfully raised. Breeders had a little more success with the Passenger Pigeon, until the birds became too inbred and produced infertile eggs.

On September 1st, 1914, at the Cincinnati Zoological Park, a Passenger Pigeon named Martha died, and with her she took all the genetic wonder that was her species. Less than four years later, the same zoo mourned the death of another bird, the planet's last Carolina Parakeet.

While Europe has seen thousands of years of civilization and countless wars, not one species of land bird is known to have become extinct. Here in North America, humans obliterated two of the most abundant in about a hundred years. Of all the extinctions for which humans can take credit, somehow these are the most senseless.

Every year, as August bears down on September, I think of those two birds, dying alone in the Cincinnati Zoo. And as I search for a way to comprehend the utter finality of their extinction, I remember the words of the naturalist William Beebe:

The beauty and genius of a work of art may be reconceived, though its first material expression be destroyed; a vanished harmony may yet again inspire the composer; but when the last individual of a race of living beings breathes no more, another heaven and another earth must pass before such a one can be again.

(William Beebe, *The Bird: Its Form and Function*, 1906,
Henry Holt and Company.)

Pages from a Birder's Diary

July 16th

The woods are pretty quiet; it's not surprising for a midsummer mid-morning. Around me, mature Douglas-firs stand, inscrutable and mute, their lives apparently in order as they move into their fourth (or perhaps fifth) centuries in this forest. Others of their kind have fallen here and there around them. Some have come down fairly recently, their thickly barked trunks now serving as major highways through the tangled undergrowth; some have been on the ground for many years, almost unrecognizable under coverings of thick moss, and the salal and red huckleberry plants that have taken root along the length of the mouldering remains.

Occasionally I hear the *ank, ank, ank* of nuthatches, and the chickadees calling to each other, as they forage through the upper storey. A Pileated Woodpecker gives his maniacal laugh from somewhere else in the forest: a voice big enough to hold its own in the serried ranks of the silent firs.

From a dense thicket off to my right comes another voice. High pitched and deceptively loud, it's the song of a Winter Wren. I turn my head to see if I can locate him. I'm in the middle of a bird census, and have been silent while I listened for birds in the forest. Guided by survey protocol, I resist the temptation to "pish" this wren in for a closer look.

A Winter Wren has an amazing song. It bubbles and trills and rolls and warbles; there can be over a hundred individual notes in its seven- or eight-second run. It is loud enough to have come from a much larger bird, just the ticket for advertising one's prowess in sound-swallowing timber like this.

The song ends and I wait. *Chip*. Wait for it. *Chip-chip*. There it is, the classic two-note Winter Wren call. The first note sounds a lot like a Fox Sparrow, but the double note is pretty hard to mistake. The next call is closer. Salal leaves move. There it is, at the base of a big fir.

It flits to a huckleberry bush growing out of an ancient nurse log directly in front of me. Any movement may spook the bird. I blink. No problem. The wren is on the log now, hopping toward me. It moves from moss to salal stem to dead twig to loose bark, changing perches perhaps once a second. I can hear the *prrrt* of its wings each time it flies. My eyes reach the end of their travel, and I have to turn my head. Still there.

These are tiny birds by any standard, but in a mature forest they seem minuscule. They are about the size of a wine cork, with a stubby little tail that is carried upright. The tail flicks regularly, and the little bird bobs its body, too, as it moves along.

The wren is now at my feet, probing here and there in the moss. It must have noticed something was different, because it is working its way up a gangly bit of salal. It stops to look at my knees, my belt, my binocular. It begins to chip again.

I stand as still as I can, but the bird knows I am here. It works its way, perch by perch, onto a dead fir sapling to my right, and now it's moving along one of the larger remaining branches, just about at eye level. Once again, my eye muscles are straining. Slowly, I turn my head. Face to face.

Winter Wrens are brown. Well, they're a little streaky, too, but mostly they're brown. At arm's length, what strikes me is the warmth of the colours. The flanks are buff, and the upperparts warm chocolate, barred with a richer brown. The eyes are black, highlighted by buffy eyebrow lines. The plumage is soft, and this bird looks to be in fine shape.

He examines me from several perches, including a look over his shoulder as he clings to the trunk of a tree to my left. He blinks, and I blink. He calls, but less frequently now. I can't decide if I am being scolded or not. Whatever the case, this tiny mite is not too concerned.

It's amazing to me that this species has evolved in the niche it occupies. It would take 8,000 of them to match me in weight, and yet they are more at home in these deep forests than I am. They are found on three continents. Most Winter Wrens are migratory, with some even making an annual round trip to Iceland. Here on the west coast, some are resident, surviving such abuse as our fickle climate may visit upon them.

The Winter Wren is known to science as *Troglodytes troglodytes*. That, translated, means "cave-dwelling cave-dweller" — a double reference to its habit of seeking heavy cover and to its domed mossy nest under a tree root. No wonder it is so at home in these woods.

The wren returns to the level of the log and continues on its way and, when it has retreated into the understorey again, I can relax and change position. The entire encounter has lasted perhaps three minutes, but in that time we had quite a little tête-à-tête, this Winter Wren and I.

I set off slowly, farther into this forest, past tall Douglas-firs and across mossy logs. I stop again, to listen for birds and, behind me, his song is there, slicing the silence, claiming these woods as his own again.

August 9th

These evening walks have much to offer. The songbirds are more active in the mornings, it's true; those are wonderful times. But while I do love being out early, I do it less often than I ought to; it's the getting there that hurts. But here, at the end of the day, is a time with a magic of its own.

The light is the way I like it best, clear and golden, and it brings the grove of Douglas-firs alive against a slate sky. By the creek, a Common Yellowthroat casts his spell, *Witchery, witchery, witch*! to keep us away. Young birds of some species or other are crashing unsteadily through the willows as they seek better cover. Pine Siskins are down, working the gleanings at the edge of the hayfield.

As we walk through a field of hay-stubble, a Barn Swallow makes a close pass. On his return, he drops below eye level, and his back catches the glow of the western sun. The feathers flash that electric blue, the colour of blued gunmetal. (I wonder: Is the swallow the colour of gunmetal, or is gunmetal the colour of the swallow?)

The swallow's passes now begin to take on a pattern, a miniature orbit of sorts, round and round us as we walk. Looking down on the subtle twists of wing and tail as it scoops up its reward, I can see that the swallow is using us, feeding on the insects kicked up by our passage.

As the sun sinks lower, my eyes are drawn higher to the golden glow in the trees. There, on a picture-perfect limb set apart from the tangle of boughs, a Red-tailed Hawk surveys the valley, creamy breast aglow with the warm light. At other times, this bird would be content to watch, secure in its lofty perch. His young, though, are probably out of the nest, and still nearby; the alarm is sounded, a sharp descending *keeeer*! The Red-tail falls away from the perch, tail ablaze in the setting sun, and banks immediately away, back into the trees, the better to guard his fledglings.

There is a low-pressure weather system moving in, and I am on the lookout for the swifts that so predictably appear with these lowering clouds. By and by, a single Black Swift makes its way across the sky in the fading light. The long slim wings look quite different from those of our Barn Swallow; its flight is less dashing and seemingly more erratic. The lone adventurer is soon swallowed by a bank of clouds almost as dark as he.

Over the breeze in the foliage, there is a faint nasal call. I listen hard, but hear nothing. Perhaps it is just the wind. Then, as we turn to walk back, there it is again. It is so familiar, and yet every year, the summer sound of the Common Nighthawk, calling overhead, sends a little shiver through me.

Along the edge of the woods now, a Purple Finch rolls off a song, hurrying to get it out before he settles for the night. Swainson's Thrushes though, comfortable in the darkest of forests, continue their ethereal ramblings as, one by one, the other birds cease. A robin gives a single note of alarm from deep in a hawthorn bush.

I stop, more out of habit than anything else, and give a little whistled trill. A couple more times, and then there is a reply: first one, and then another, a pair of Screech-Owls. The night shift has just come on.

This walk ends in darkness, the light and sound and coolness on our skin different from when we set out. I think of the owls, active through the night, retiring at first light to make way for the bustle of the early risers. I really should get out for some early morning birding soon, but for now, this will do just fine.

Afterword

I would like to leave readers on an "up" note, with a feeling that, despite some occasional bad news, in the end our birds will be okay, that we will be able to enjoy them in the way we always have. I wish I could be certain of that.

To be sure, there are success stories. Trumpeter Swans have returned from near-extinction, and the West Coast population of Purple Martins is also recovering, thanks to a major effort on the part of volunteers. There is some solid evidence of increases in other species, but the truth is, even the scientists acknowledge that they don't know why it's happening. On the other hand, many more species are in decline.

In my own time as a birder, I have witnessed the disappearance of the Western Bluebird from Vancouver Island. Despite the valiant efforts of Harold Pollock and his volunteers, who put up over 1,200 nestboxes, the bluebirds slipped past the invisible threshold and vanished. I have spoken to other people who remember when Western Meadowlarks and Lewis's Woodpeckers nested here, too. They are gone now, and the Vancouver Island race of the Vesper Sparrow is reduced to probably only a few birds.

Elsewhere in the province, we have lost the B.C. populations of Sage Grouse and Yellow-billed Cuckoo. The Burrowing Owl, also considered extirpated in B.C., may be recovering as a result of attempts at reintroduction.

Still, September has come around one more time, and so have all the birds it brings with it. The Turkey Vultures are piling up on southern Vancouver Island,

and a new generation of shorebirds is on the beaches and mudflats, every young bird making its own way south for the first time. Golden-crowned Sparrows are turning up at feeders again, back to spend the winter with us. It's a time filled with the promise of another year, a wondrous confirmation that these tough and resilient birds have survived, in the face of odds that would daunt us humans. Every year I take some comfort in this, and every year I remind myself: Don't take this for granted.

I give myself this admonition because my own personal (and I hasten to add, unscientific) observation is that birds are unequivocally declining in numbers. It is not just my imagination, though. Birds today face ever greater challenges. But this fact has not gone unnoticed, and the good news is that a lot of people are not taking our birds for granted.

Despite overwhelming cuts by one government after another to already small environment budgets, scientists still manage to do some research that will help us understand how to sustain bird populations. Non-profit conservation organizations have tried to take up some of the slack, protecting habitat and fostering better land stewardship, despite the fact that they must work with only about 3 percent of all charitable donations.

Volunteers make a tremendous contribution, working at bird-banding stations, wildlife recovery facilities and nature centres. Amateur naturalists put in thousands of hours in a host of "citizen science" programs, like bird censuses, nest-monitoring programs and Christmas Bird Counts.

Backyard birdwatchers play a role by "naturescaping" their properties, and enhancing their gardens for wildlife. Cat owners who keep Fluffy and Muffy inside help to save the lives of millions of birds.

It gives me hope to see the commitment made by everyday bird lovers, a hope that we can, little by little, make a difference in the lives of the birds we come to know. The effort is made, I think, because people have felt a connection with birds.

I recall leading a birding course one fall, and the group was walking across a fallow agricultural field. It was a splendid morning, with Short-eared Owls coursing over the hay-stubble, and wintering Western Meadowlarks flashing egg-yolk breasts in the autumn sun. The group was ebullient about this wonderful new spot I had introduced them to. Then, when I explained that the field was slated to become a golf course, they were stunned. They understood immediately what was at stake, and all it had taken was for them to be there, to see the birds, to connect with them in their daily lives.

Most of us probably do not experience that connection with birds in quite so immediate a way, but we all feel it. It's there, in the bright eye of a chickadee, or the immaculate plumage of a Cedar Waxwing, or perhaps the midnight duet of a pair of Great Horned Owls. And it is this connection, I think, that holds the hope of a future for the birds. It is our growing recognition that birds have a right to a future. And that future is not theirs alone — it is our future too.

We are not just passive observers of birds, as aesthetic adornments in our lives. We are partners with them, voyagers on the same ship, the celestial vessel Planet Earth, and we are all bound for the same future. Every troubled passage our ship enters holds danger for us all. Every time one of the voyagers is stricken on board, it is a danger signal to all the other travellers, too — a sign that all is not well with the ship. The canary in the mine lives on this ship, too — in fact as well as in metaphor.

So birding for me has become a more complicated game than it used to be. I look at birds now with a different eye — not a fearful eye, exactly, but a watchful eye. When I see a young fall warbler feeding outside my window, I know it is putting on fat to sustain itself en route to its winter destination. It has never been there, but it knows where it is going. What it cannot know, and the question I ask, is will "there" even be there, when it gets there? And then, with every spring, I watch the birds again, not so much to see when they return, but if they return. As long as they arrive each spring and fall, I know the birds are okay. And as long as they're okay, maybe we're still okay too. There is no doubt that this is a different way of looking at birds.

If this changing reality has affected the way I watch birds, it has not altered the underlying way that birds cast their spells on me. I still smile to see the electric blue of a Mountain Bluebird, and thrill to a jaeger's turn of speed. My spine still tingles at the hollow-barrel calls of a Barred Owl.

And watching now, as a Red-tail slips out over the Strait of Juan de Fuca, I feel a familiar great peace. The hawk has hit the cold marine air but it hesitates only briefly, and then commits to the crossing, bound by an unknown past, and flying into an uncertain future. Once again, I am left in quiet awe, and that part will never change. No, there is no need for admonition: I cannot — I will not — take this for granted.

Appendix
Tools of the Trade: Binoculars

Virtually all birders use binoculars and, as a group, they use binoculars more than anyone else. They expect a lot of their binoculars, and they depend on them to perform well under a wide range of conditions. If birders are to get the most out of their binoculars, it makes sense that they should know something about them.

The binocular (it is singular; not a "set" or "pair" of binoculars) is descended from the field glass, which is essentially two telescopes fastened together. The field glass was a remarkable advancement, but it was limited to low magnification.

An Italian by the name of Porro determined that if prisms were employed in a field glass to bend light in a folded path, greater magnification could be achieved with no increase in size. Today, the majority of binoculars use the prism system that bears his name. Porro prism binoculars have the familiar binocular shape, with objective lenses (at the front) set wider apart than the eyepieces. Some have bodies that are cast in one piece and others are assembled from several pieces; there is no difference in performance.

The other prism design in use today is the so-called "*dach*" or "roof" prism. There are several such designs, but most are so named because one of the prism faces is ground to look like a gable roof instead of being flat. Roof prism binoculars normally look like a letter H; usually the light path is folded inside so that it appears to enter and leave the binocular in the same line.

There is a lot of debate about which prism design is better but it boils down to personal preference. Roof prism binoculars are more expensive to make; porros give the best value for money. There is considerable overlap in size, weight, optical quality and performance.

On any binocular, you'll see brand names and model names, which don't tell you much about the binocular. More important are the numbers you'll find. Let's look at a typical binocular. First of all, we see the designation: 7 x 35. The first number says that the binocular magnifies the subject seven times (7x); a Peregrine on a distant telephone pole will look seven times bigger. The second number is the diameter in millimetres of the objective lens (the big end). In our sample, it is 35 millimetres across.

The larger this lens is, the more light it lets into the binocular. But the larger it is, the larger (and heavier) the binocular is too. We're going to have to compromise, so how much light do we need? In daylight the pupils of our eyes are normally dilated to a diameter of about 2.5 or 3 millimetres. But in low light, for example at dawn or dusk, or looking for quetzals in a Central American rainforest, our pupils can expand to as much as 7 or 8 millimetres.

We know that the objective is letting in a shaft of light that's 35 millimetres in diameter. To find out how much light reaches the eye, divide the diameter of the objective lens by the magnification: 35 divided by 7 is 5; this binocular is delivering a 5-millimetre shaft of light to each eye. You can see this on your own binocular if you look into the eyepiece from a distance of a foot or two; the circle of light you see is called the "exit pupil," and it's one of the most important things you need to know about binoculars.

In a 7 x 20 compact binocular, the magnification is the same (7x), but the exit pupil is a little less than 3 millimetres; just enough to flood the pupil with light on a sunny day but with no leeway when the light is poor. There is another point to consider, too. Trying to line up your pupils on the two images formed by your binocular is much easier when the binocular's exit pupil is larger than the pupils of your eyes. This is one reason that 7 x 50s are easier to use on a moving boat.

The next number that concerns us tells us how wide the field of view is. It's written on the binocular in one of two ways. The first is in degrees: our sample binocular has a field of view that covers 9.3 degrees of a circle. Sometimes the field is given as the width in feet of the image seen at a distance of 1,000 yards; our binocular would offer about 500 feet at 1,000 yards. For purposes of comparison, one degree is approximately equal to 52.5 feet at 1,000 yards.

From here, binocular performance becomes a matter of compromise. You may gain one feature but you will lose another. No matter what anybody tells you, there is no one best binocular for birding, but there is a best binocular for you.

For example, while 7-power binoculars were the order of the day some years ago, today most birders use 8s or 10s. Greater magnification means bigger birds, yes, but at what cost? As the power goes up, remember that brightness and field of view are reduced. Consider another factor: If you're getting your first look at a Western Tanager, a 10x binocular will also be magnifying every beat of your accelerated pulse, and every muscle tremor from the hike up the hill. You may have a beautiful but shaky tanager. I'd suggest 8 power as a good place to start, with 9 or 10 having merit only if you're certain you can hold them steady comfortably. Try reading a typed page at a distance of 8 or 10 metres with a 10x binocular and you'll see what I mean.

The next feature to look at is objective size. The higher the magnification of the binocular, the smaller the exit pupil will be, unless you choose a larger objective, which will make the binocular larger and heavier.

I recommend an exit pupil of at least 4 millimetres for birding. An 8 x 32 or a 10 x 40 meets this requirement. If you are prepared to carry more glass, you can go higher. If you want to reduce the weight, you may find a compact model which will suit you, but you will at least occasionally find yourself short of light.

There is another important factor that affects brightness in a binocular. Every time a beam of light passes from air to glass, or from glass to air, about 5 percent of it is reflected away or scattered inside the binocular. In an average binocular, with many refracting surfaces, the total light loss can easily be as high as 40 percent. To reduce this loss, ultra-thin coatings, originally of magnesium fluoride, were applied to the refracting surfaces. This resulted in a light transmittance of something over 80 percent.

Better results are achieved if the coating consists of several thin layers of different compounds; transmittance climbs to over 90 percent, and some manufacturers claim 95 to 97 percent transmittance. This advancement is known as "multicoating," and most of the better binoculars today incorporate this feature.

There are different opinions about the optimum field of view in a birding binocular. One school argues that a wide field of view makes it easier to track fast-moving birds. The other maintains that it's better to put all your glass on the subject. Binoculars which do not have wide fields have fewer lenses in them and are thus lighter. They also show greater sharpness and less distortion at the edges of the field. In any case, width of field decreases as magnification increases, so be sure to consider this fact in your "best binocular" equation.

It is very important that birding binoculars be able to focus on birds as close as possible. Many binoculars will focus no closer than 8 or 10 metres. I suggest 3 to 4 metres as a minimum; some binoculars will focus down to less than 2 metres —

they're terrific for butterflies too. Try them yourself, because this distance varies among individual users.

Eyeglass wearers do not see the full field of view in a standard binocular. Eyecups which pop down, twist down or fold down are helpful; they allow the binocular to sit closer to the eye, so the eye sees a fuller field of view. Better still are the "long eye relief" models, which give eyeglass wearers a full field of view even with glasses on.

Most binoculars are not waterproof. Truly waterproof binoculars are sealed and filled with nitrogen; they are very useful for birding in a wide range of extreme conditions, but repairs are costly, and must be done at an authorized facility where resealing and nitrogen purging can be done.

Next we get into an area where personal preferences and individual requirements come into play. Large hands may be uncomfortable with a small binocular. Long eyelashes may require deeper eyecups. Narrow eye-spacing will eliminate many models. Some birders prefer fast focussing, while others want the precision of a slower mechanism.

The real test is using the binocular. Tack-sharpness is little consolation if the binocular is just not comfortable. My suggestion is that you narrow the choices to several models which come close to meeting your criteria and then base your decision on which feels the best. It's better to compromise a little on one feature or another than to saddle yourself with a theoretically ideal binocular that feels foreign every time you pick it up.

Finally, a list of dubious features, and reasons to consider avoiding them. Fast-focus mechanisms are imprecise, and can even be affected by pressing the binocular to your face. Permanently focussed binoculars (with no adjustment) will not focus closer than 12 to 15 metres and are useless for birding. Zoom binoculars are fraught with limitations: At higher magnifications, there is too little light, and they cannot be hand-held. Extra glass means lower resolution and greater light loss. More moving parts means less precision and less optical quality for your dollar. Stay away from any of these binoculars. If you currently own one, learn to minimize its limitations until you can retire it.

You'll do well to buy the best that you can afford, but you can maximize the value you are getting for your dollar by checking the various models for the performance features you want. Once you've made a choice, your binocular will soon become an old friend, and you can get down to the business of finding and enjoying birds.

Selected References

There are thousands of bird books, some excellent and more than a few a waste of paper. These are the ones that stand out; the books I think of as "first off the shelf" when I need information.

Field Guides

There is no "best" guide. They all have strengths and weaknesses, and it is not overkill to own several. Artists' illustrations are much preferred by experienced birders, but photographs are sometimes useful in more advanced guides to bird families, like shorebirds. The following are the guides used by most birders; others have their place too.

Baughman, Mel, ed. 2004. *Field Guide to the Birds of North America.* Washington: National Geographic Society. Revised regularly since 1983.

Peterson, Roger Tory, ed. 1990. *A Field Guide to Western Birds.* Boston: Houghton Mifflin. This is still my favourite guide, with the best range maps of all the guides.

Sibley, David Allen. 2000. *The Sibley Guide to Birds.* New York: Alfred A. Knopf. It is large — handbook size — and expensive. There are now eastern and western guides which are pocket-sized and very good, but I find the maps more confusing than those in the larger guide.

Sibley, David Allen. 2002. *Sibley's Birding Basics.* New York: Alfred A. Knopf. This is not a field guide per se, but a lesson in bird identification. Sibley illustrates how birds can appear different at different ages, in different lights, and with fresh or faded plumage.

Family Accounts

These guides go into greater detail on families of birds. There are excellent family accounts for other groups as well, such as warblers or sparrows, but the standard guides cover those adequately.

Dunne, Pete, David Sibley and Clay Sutton. 1988. *Hawks in Flight.* Boston: Houghton Mifflin. One of the best bird books ever written.

Grant, P. J. 1986. *Gulls: A Guide to Identification.* Vermilion: Buteo Books.

Harrison, Peter. 1983. *Seabirds: An Identification Guide.* Boston: Houghton Mifflin.

Paulson, Dennis. 1993. *Shorebirds of the Pacific Northwest.* Vancouver: University of British Columbia Press.

Wheeler, Brian K. and William S. Clark. 1995. *Photographic Guide to Raptors of North America.* Princeton: Princeton University Press.

Regional Guides
Campbell, R. Wayne, Neil K. Dawe, Ian McTaggart-Cowan, John M. Cooper, Gary W. Kaiser, Andrew C. Stewart, Michael C. E. McNall and G. E. John Smith. 1990-2001. *The Birds of British Columbia*. vols. I-IV. Victoria: Royal British Columbia Museum. Vancouver: University of British Columbia Press.

Cannings, Robert A., Richard J. Cannings and Sydney G. Cannings. 1987. *Birds of the Okanagan Valley*, British Columbia. Victoria: Royal British Columbia Museum.

Godfrey, W. Earl. 1986. *The Birds of Canada*. Ottawa: National Museum of Canada.

General Books About Birds
Baicich, Paul J. and Colin J. O. Harrison. 1997: *A Guide to the Nests, Eggs and Nestlings of North American Birds*. Princeton: Princeton University Press.

Bodsworth, Fred. 1954. *The Last of the Curlews*. Toronto: McClelland & Stewart. A fictional account of the migration of an Eskimo Curlew, a species now possibly extinct.

Choate, Ernest A. 1985. *The Dictionary of American Bird Names*. Boston: The Harvard Common Press.

Elphick, Chris, John B. Dunning Jr. and David Allen Sibley. 2002. *The Sibley Guide to Bird Life and Bird Behaviour*. New York: Alfred A. Knopf.

Ehrlich, Paul R., David S. Dobkin and Darryl Wheye. 1988. *The Birders' Handbook*. New York: Simon & Schuster. One of my most-used references.

Mearns, Barbara and Richard. 1992. *Audubon to Xantus*. San Diego: Academic Press. I drew heavily on this book for some of the accounts of early ornithologists.

Quammen, David. 1996. *The Song of the Dodo*. New York: Simon & Schuster. This book inspired the essay "In the Geological Wink of an Eye."

Terborgh, John. 1989. *Where Have All the Birds Gone?* Princeton: Princeton University Press.

Terres, John K. 1980. *The Audubon Society Encyclopedia of North American Birds*. New York: Alfred A. Knopf. A remarkable book, especially for a single author.

Bird-finding Guides
I've only included titles for British Columbia and Washington; there are hundreds more.
Aitchison, Catherine J., ed. 2001. *The Birder's Guide to Vancouver and the Lower Mainland*. Vancouver: Vancouver Natural History Society and Whitecap Books.

Opperman, Hal. 2003. *A Birder's Guide to Washington*. Colorado Springs: American Birding Association. One of an extensive series of guides by the ABA.

Taylor, Keith. 1998. *The Birder's Guide to British Columbia*. Vancouver: Steller Press. Note that this guide does not cover Vancouver Island.

Taylor, Keith. 2000. *The Birder's Guide to Vancouver Island*. Vancouver: Steller Press.

Binoculars

Armstrong, Alan. 1990. *Binoculars for Birders*. Madison: Avian Press.

Hale, Alan R. 1991. *How to Choose Binoculars*. Redondo Beach: C & A Publishing.

Gardening for Wildlife

Campbell, Susan and Sylvia Pincott. 1995. *Naturescape British Columbia*. Victoria: Ministry of Environment, Lands and Parks. The Naturescape set includes three booklets: a provincial guidebook and two guides for each region of British Columbia.

Merilees, Bill. 2000. *The New Gardening for Wildlife*. Vancouver: Whitecap Books.

Pojar, Jim and Andy MacKinnon. 1994. *Plants of Coastal British Columbia*. Vancouver: Lone Pine Publishing. Probably the best field guide to plants of the west coast.

Pettinger, April and Brenda Costanzo. 2003. *Native Plants in the Coastal Garden*. Vancouver: Whitecap Books.

General Natural History Books

Butler, Robert. 2003. *The Jade Coast*. Toronto: Key Porter Books.

Cannings, Richard and Sydney Cannings. 1996. *British Columbia: A Natural History*. Vancouver: Douglas & McIntyre.

Penn, Briony. 1999. *A Year on the Wild Side*. Victoria: Horsdal & Schubart.

Weston, J. and D. Stirling. 1986. *The Naturalist's Guide to the Victoria Region*. Victoria: Victoria Natural History Society.

Yorath, C. J. and H. W. Nasmith. 1995. *The Geology of Southern Vancouver Island*. Victoria: Orca Book Publishers.

Index of Bird Species